**Plasticity and Pathology**

Berkeley Forum in the Humanities

# Plasticity and Pathology
## On the Formation
## of the Neural Subject

## Edited by David Bates
## and Nima Bassiri

Townsend Center for the Humanities
University of California, Berkeley

Fordham University Press
New York

Library of Congress Cataloging-in-Publication Data

  Plasticity and pathology : on the formation of the neural subject / edited by David Bates and Nima Bassiri. — First edition.
      p.   cm. — (Berkeley forum in the humanities)
  The essays collected here were presented at the workshop Plasticity and Pathology: History and Theory of Neural Subjects at the Doreen B. Townsend Center for the Humanities at the University of California, Berkeley.
  Includes bibliographical references and index.
  ISBN 978-0-8232-6613-5 (cloth : alk. paper)
  ISBN 978-0-8232-6614-2 (pbk. : alk. paper)
  I. Bates, David William, editor. II. Bassiri, Nima, editor. III. Doreen B. Townsend Center for the Humanities, sponsoring body. IV. Title. V. Series: Berkeley forum in the humanities.
  [DNLM: 1. Neuronal Plasticity—Congresses. 2. Brain—physiopathology—Congresses. 3. Synaptic Transmission—Congresses. WL 102]
  QP355.2
  612.8'1—dc23
                                                                    2015026932

Printed in the United States of America

18 17 16    5 4 3 2 1

First edition

# Contents

# Illustrations

## Contributors

NIMA BASSIRI is Collegiate Assistant Professor and Harper Schmidt Fellow at the University of Chicago.

DAVID BATES is a professor and chair of the Department of Rhetoric at the University of California, Berkeley.

JOSEPH DUMIT is a professor of anthropology and director of Science and Technology Studies at the University of California, Davis.

CATHY GERE is a professor of history at the University of California, San Diego.

STEFANOS GEROULANOS is an associate professor of history at New York University.

KATJA GUENTHER is an assistant professor of history at Princeton University.

CATHERINE MALABOU is a professor of philosophy at the Centre for Research in Modern European Philosophy at Kingston University and a professor of philosophy at the European Graduate School.

EMILY MARTIN is a professor of anthropology at New York University.

TODD MEYERS is an assistant professor of anthropology at Wayne State University.

HANNAH PROCTOR is a graduate student in English and humanities at Birkbeck College, University of London.

TOBIAS REES is an assistant professor of anthropology and social studies of medicine at McGill University.

LAURA SALISBURY is a senior lecturer in medicine and English literature at Exeter University.

*David Bates and Nima Bassiri*

# Preface

THE ESSAYS COLLECTED here were presented at the workshop Plasticity and Pathology: History and Theory of Neural Subjects at the Doreen B. Townsend Center for the Humanities at the University of California, Berkeley. As co-organizers of this event, we were trying to put together several different strands of scholarship that have taken up the challenge of critically engaging with contemporary neuroscience research. As the neurosciences have gained considerable popular interest in the past decade or so, the impact of this work on the humanities and social sciences has been less clear. While some have embraced with great enthusiasm the findings of neuroscience and used them in their own research, others willfully ignore the field altogether despite the obvious relevance to anyone working on the way human beings think, act, and are fashioned in the world. This volume is an effort to explore how scholars in the humanities and social sciences might begin to think critically about the historical and conceptual (not to mention institutional) development of the human subject in the age of neural science, as a way of raising questions about who we are and who we might be.

We have chosen the intertwined fields of plasticity and pathology as our starting point because they reveal most clearly that the history of neuroscience is hardly one of strict reductionism. Both of these conceptual fields are and have been highly ambivalent, suggesting in their own way the degree to which the nervous system—and especially its central organ, the brain—has often been considered an ever-evolving, dynamic, and transformative space. Plasticity is essential to, for example, the developmental potential of the infant brain and therefore an essential dimension of the human self. Plasticity also marks the possibility of a radical change. It is precisely this flexibility that helps the nervous system respond to injury and pathological conditions. And yet to transform the brain so radically—in reaction to injury or through internal transformations—is to transform the subject itself, to make in a way a new human being. Therefore, rather than trying to stabilize the concepts of plasticity and pathology as definitive categories, we actually want to emphasize the underlying fluidity at the heart of these concepts—not just according to their contemporary prominence in neuroscience but because they also index a set of important conceptual developments and possibilities in the history, anthropology, and philosophy of mind and brain medicine.

The scholars collected here represent different ways of approaching the question of the neural foundations of a human subject. The essays range from anthropological accounts of subject formation, investigations of contemporary neuroscientific research, and historical analyses of key theoretical debates in the formative decades of neuroscience to forays into artificial nervous systems with their own artificial forms of plasticity, narrative interpretations of neural subjects, and philosophical reflection on the nature of the mind. They are intended to display the conceptual variability of these terms and to indicate a way of thinking about their varied historical development and their continued transformation in the present and into the future. Plasticity and pathology, as neurological concepts, point to complicated phenomena in the history and theory of the human sciences. To grasp the significance of these phenomena we need an open and multidisciplinary approach.

WE WOULD LIKE to thank Alan Tansman, director of the Townsend Center, for his intellectual and material support for this project, which is part of a broader initiative on "Thinking the Self." We would also like to thank Anthony Cascardi, dean of arts and humanities, and Carla Hesse, dean of social science, as well as the Department of Rhetoric for generously funding the original workshop. We are indebted to Teresa Stojkov, associate director of the Townsend Center, for all her help making this edited collection a reality.

*Emily Martin*

# 1 Toward an Ethnography of Experimental Psychology

AN ENDURING QUESTION in the history and philosophy of science is: What do we mean by *objectivity* and *subjectivity*?[1] In their historical overview Lorraine Daston and Peter Galison set out three phases of scientific knowledge over the centuries, from "truth to nature," to "mechanical objectivity," to "trained judgment." On the one hand, the epistemic virtue of "mechanical objectivity" strives to "capture nature" while eliminating any intervention on the part of the researcher. Exemplary photos of snowflakes show how they were deliberately presented to retain asymmetries that show the capture of nature with as little interference from the researcher as possible. The epistemic virtue of "trained judgment," on the other hand, is one in which intuitive or aesthetic elements can enter into how a researcher interprets a brain image, for example. One of their examples of trained judgment is, in Daston's words, an "image of the magnetic field of the sun [mixing] the output of sophisticated equipment with a 'subjective' smoothing of data— the authors deemed the intervention necessary to remove instrumental artifacts."[2] Epistemic virtues are not rigidly *separated* into different regimes as much as they are *characteristic* of them. In this paper I offer a modest engagement with the issue of "training" in relation to psychological experimentation as a way of understanding how experimental psychologists grapple with issues of subjectivity and objectivity. We will see that a basic goal of the experimental method in cognitive psychology is to keep the human subject stable in time and space. Since human beings are, in many other contexts in the history of neurology, understood as constitutively mobile, changeable, and profoundly plastic, this is a hard job. We should not be surprised if accomplishing such a hard, perhaps impossible job produces some contradictions.

In what follows I turn to the history of experimental psychology, which was closely connected to anthropology at its beginning, to understand what role subjectivity and objectivity were granted at an earlier time and to offer a partial solution to the puzzle I will introduce in due course. I will describe two incidents from my current fieldwork with experimental cognitive psychologists (ongoing in three labs in the United States) to present a puzzle about what role subjectivity has today in experimental psychology. I will relate what experimental cognitive psychologists explained to me as the ways they maintain objectivity in their experiments through various forms of control.

## Incident #1

DURING THE TWO-YEAR wait I endured while trying to obtain permission to do fieldwork in experimental psychology labs, I passed the time by volunteering to be a subject through public requests posted by academic psychology departments on the web or on bulletin boards. I was struck by how irrelevant my experience as a subject was to the experimenters. In one experiment, for example, I was hooked up with electrodes used to measure small facial movements of which I was unaware. These would indicate my emotional responses to photographs presented on the computer screen in front of me. I pressed keys on the keyboard to register my conscious responses to these photographs. A software program tallied the results. My responses were produced, I was told, by specific parts of my brain. What the researchers sought were data about how my *brain* reacted to the photographs. Whatever was going on in my conscious experience could be ignored. For example, although the monitor I was to attend to and make my responses to was right in front of me, just on my left was another monitor, which would show the varying electrical impulses from my electrodes. I noted to the experimenter that I could easily see the read-out of my own responses, and she said, "That's fine, it doesn't matter." It was as if the experimental setting were considered such a secure enclosure that it could render minor aspects of my environment insignificant, including even my ability to see the readout from my own facial electrodes.

## Incident #2

A GRAD STUDENT in a different lab told me about an experiment he was thinking of doing that would build on earlier research. He said, "The earlier experiment didn't work, and one of the reasons that it didn't work is they didn't train people. They just put them immediately in the scanner and tried to do everything at that time." Coincidentally, in the lab meeting for yet another lab that same week there was a long discussion of "test-enhanced learning." I learned that researchers have found experimentally that any form of practice before a test improves outcomes. It does not matter whether the student gives wrong answers or is given no questions to answer at all but only told to mechanically fill in the answer bubbles: practice of any kind before a test improves performance.

At this moment, sitting in the lab meeting, I was shocked. The grad student had just told me that lack of practice before an experiment was a factor in the failure of an experiment, but in the lab meeting I had just learned that psychological research had demonstrated the power of the "practice effect" in impinging on performance. Only then did I realize that in almost all the experiments in which I had volunteered as a subject I was being trained before the experiment started. I certainly remembered the practice exercises. But I had thought of them as if I were a student who was being tested about whether I was an adequate subject, not as a form of training that would affect my performance in the experiment. After the practices I couldn't help asking, "How did I do?" None of the experimenters ever answered that question except to say "Fine" with a smile, as if my question really didn't have an answer. Nor were my experiences unusual: the dozens of textbooks on experimental method I have consulted recommend the use of practice trials to "stabilize the subjects" "before the experimental conditions of interest are introduced."[3]

So the puzzle conveyed by these incidents is this: in the first case the experimental setting is held to be so powerful that it can render aspects of the subject's experience irrelevant. In the second case the experimental setting actually includes training techniques that deliberately affect the subject's experience. These aspects of the two settings seem to work in opposite directions. Another way to put this is this: in the first incident it is the subject's experience that is at issue; in the second it is both the subject's experience *and* the experimenter's experience, in the sense that the experimenter's goal of achieving an *effect* might be said to have intervened in the experiment.

Understanding why I was told that my experience of seeing the readout of my own responses didn't matter comes fairly easily, thanks to historians of psychology (such as Betty Bayer, Kurt Danziger, and Jill Morawski) who have described how "introspection"—the role of the subject's own experience (his or her subjectivity)—largely came to be ruled out of experimental settings by the mid-twentieth century. According to John Carson, by eliminating introspection, the human subject was thus "transformed into a usable experimental object."[4] Similarly historian Kurt Danziger has remarked on the progressive elimination of the subject's own *experience* from psychology and pointed out that where the effort has been made to reintroduce it, the refusal has been absolutely relentless.[5] Daston and Galison say that *subjectivity* has at times been a "fighting word."[6]

So far so good. But if the subject's experience can be ignored, how do we understand the need for *training* subjects? Psychologists generally certainly accept the virtue Daston and Galison describe as "mechanical objectivity," "capturing nature" while avoiding interventions by the researcher. But one kind of intervention is apparently an exception: experimenters design training for subjects that happens routinely *as a part of the experimental protocol*. We could ask whether another of Daston and Galison's virtues—"trained judgment"—is playing a role here, but we would have to realize an important difference: for Daston and Galison, it is the *experimenter* whose judgment is trained; in my fieldwork labs it is the *human subject* who is trained. How can a subject who is *trained by the experimenter* play a part in producing (objective) scientific knowledge?

### The Experiment

TO MOVE TOWARD the beginning of an answer, we might start with the concept and practice of the *experiment* in psychology labs. I have learned in my psychology classes and labs that good experiments need to have the following characteristics. Dependent variables are distinguished from independent variables. All dependent variables should be clearly measurable. *Dependent variables* are commonly measured by reaction time, the interval between the time the stimulus is presented and the time the subject presses a key. Commonly these are behaviors in a task, such as distinguishing between different stimuli or remembering a stimulus from earlier in the experiment. The experimenter introduces *independent variables* and controls them precisely. These are called manipulations because the independent variables are meant to cause a change in the dependent variables.

I have been told this is difficult to follow, and it is for me too. Imagine a simple scene in an experiment. The subject sits before a keyboard and monitor. Stimuli are flashed on the screen, and the subject presses designated keys to indicate her response: *yes,* it is a word, or *no,* it is not a word is a simple example. Then the numerical results are logged on a spreadsheet. The reaction time between stimulus and response is the dependent variable and a small number of conditions are the independent variables, chosen by the researcher. These are the simple elements of a behavioral experiment, but they are also the basis of experiments using more elaborate technological methods, such as EEG measurements on the scalp or brain imaging in an fMRI machine. When all this is in place, the experimenter may see an *effect.* An effect is demonstrated when subjects produce statistically significant different reaction times under different conditions. Psychologists say some colleagues are especially good at "getting effects." Others are not so lucky.

The subjective experience of those sitting in the subject's chair is not supposed to be involved. The processes being measured (cognition) are not knowable to the subject—under ordinary circumstances we do not know what we remember best, what we react to most emotionally, or how we respond to a manipulation involving risk or failure. And we certainly do not know what parts of the brain may be processing these cognitive operations.

## Stabilizing the Subject in Time and Space

YOU CAN SEE that *control* of variables plays an important role in an experiment. But so does control of the *subject* as a living, human person. A key part of a psychology experiment is stabilizing the subject in time and space. Historically a simple method of holding the subject still in space was the bite bar, dating to experiments by the physicist Hermann von Helmholtz and reiterated in twentieth-century American college psychology labs.[7] Its contemporary descendant is the fixation point. The fixation point is an image of a cross, in the shape of a plus sign that appears on the computer monitor in an experiment, usually between trials. No one but me thinks it is interesting! When I ask about it, people say, "Oh, it is just to prevent subjects from looking around all over the place."

In the dozens and dozens of experiments in which I have been a subject, I've noted that experimenters do not generally explicitly say to subjects, "Look at the fixation point when it appears." (I heard something like this only once.) One time, in an ongoing experiment I was observing, I asked the graduate student what the fixation point was for. Only then did he tell the subject to look at the fixation point.

Perhaps one reason subjects don't need to be told about the fixation point is that there is another way a living human being can be held steady in time and space so that comparable data can be extracted from him or her. This modest stabilizing technology, central to the psychological experiment, is the table. Obvious and overlooked, the table is nonetheless an essential accompaniment to civilized human life: the first thing Robinson Crusoe did after being shipwrecked on his island was build a table.

A table is a technology that stabilizes people and things in space for a time. The table, with its chair, enforces a posture of attention to what is on it. It permits display and use of other tools and enables precise recording on paper. It also allows the display of disparate materials on the same plane in space. Bruno Latour explained the effect of this as he watched botanists in the field arranging soil and plant samples on tables: "Specimens from different locations and times become contemporaries of one another on the flat table, all visible under the same unifying gaze."[8] The flat plane provided by the table enables the abstraction of dissimilar specimens into categories.

Open and inviting a table might seem, but once you are sitting at it, certain forms of courtesy might serve to hold you there. If the table can be thought of as a kind of trap (following Alfred Gell) to capture and contain a subject, it is a disarming one—it looks so placid and innocent for something that has the potential for powerful constraint.[9] The table is so embedded in the experimental context that it escapes notice, even though without it the stability of the subject in space and over time would be difficult if not impossible to achieve. Once it becomes evident that the table is an active artifact in the production of knowledge, new possibilities for opening up the nature of the experimental space in psychology abound. Latour was right to say that "laboratories are excellent sites in which to understand the production of certainty, [but] . . . they have the major disadvantage of relying on the indefinite sedimentation of other disciplines, instruments, languages, and practices. . . . In the laboratory there is always a pre-constructed universe that is miraculously similar to that of the sciences."[10] After a discussion of the table's role in experiments, one of my researcher interlocutors began puzzling about what it would take to conduct an experiment about, say, memory in a crowded coffee shop instead of an experimental setting. This was disconcerting to him because leaving the laboratory would mean leaving a world of tables, flat, one-dimensional, and still. But anthropologists should take note: even the busiest coffee shop has its tables too.

In my current project tables are ubiquitous. Tables, with their chairs, keep one's body in place. In all the experiments I participated in, the experimenter made frequent and repeated requests concerning tables: Sit here at the table. Pull your chair closer to the table. Put your hand on the table. Rest your hand flat on the table. Arrange the keyboard conveniently on the table, etc. And of course tables hold computers, monitors, keyboards, and recording equipment steady. In the contemporary lab the place of the psychological subject in relation to the equipment is not open for debate. The subject sits at a table and yields data to the machines. You might say that the fixation point is ancillary to the table.

Today the psychological experiment seems governed by controls that make human subjects into "data-emitting machines" whose experience beyond what is controlled is irrelevant. We still do not have a way to understand the logic of practice or training sessions right before the data are collected. Remember that Daston and Galison's "mechanical objectivity" rules out intervention in the capture of nature on the part of the *researcher*. "Trained judgment" (a more recent epistemological virtue) allows training of the *researcher's* judgment, but this amounts to training in how to see "nature." If "nature" takes the form of a subject's psychological processing, the subject's experience falls through the cracks.

## Wundt's Lab

SO, IN SEARCH of an answer, I now turn back in time, as I promised, to the history of experimental psychology to understand what role the experience of the subject had in the nineteenth and early twentieth century and to offer a partial solution to the puzzle I've raised. I begin with the psychological experiment in the German psychology lab of Wilhelm Wundt in Leipzig, the first of its kind in the world. Historians like Ruth Benschop and Deborah Coon helped me to understand the technologies that enabled precise measurement of time intervals in Wundt's Leipzig laboratory. As Coon explains, laboratory hardware standardized and regulated the physical stimuli to which the subject would respond, and "it also gave quantified, standardized output [while using] the introspective method."[11] The subject was to record both when he perceived the stimulus and when he *recognized its meaning*. This was intended to capture *conscious* experience.

Measuring introspective responses did not imply vagueness! In the Leipzig lab, Benschop and Draaisma write, "a veritable culture of precision emerged, in particular in the area of the measurement of reaction times to sensory stimuli."[12] Immense effort went into the technology of precise and exact measurement of time. Of course this required new technology, which researchers produced by the cartload.[13]

Perhaps even more important, the subject himself had to be *standardized*. As Coon explains, even though, "in the early stages of psychology's development, typical experimental subjects were professors and graduate students, not experimentally naive college sophomores and white rats," there was still "too much individual variation among these flesh-and-bone introspecting instruments. In order to standardize themselves as experimental observers, therefore, psychologists resorted to long and rigorous introspective training periods. . . . Only if *introspectors themselves were standardized* could they become interchangeable parts in the production of scientific psychological knowledge." Wundt "insisted no observer who had performed less than 10,000 of these introspectively controlled reactions was suitable to provide data for published research."[14] What made the reaction times they were measuring "introspective" is that the subject was instructed to consider his introspective experience in deciding when he had recognized the stimulus. The training, the practice sessions, were required to make sure everyone in the lab was introspecting in the same way and producing reaction times within a similar range.

Standardization also extended to "regularity *outside* the context of experimental practice." James Cattell, a student of Wundt's, relates that he followed a strict scheme of physical exercise, prescribed by Wundt.[15] He remarks in a letter to his parents that the students were required to walk three to six miles a day.[16] Wundt and his collaborators aimed at measuring processes in what has been called "the generalized mind," those parts of mental life shared by all human adults alike. As Benschop explains, "Being practised in appearing in experiments helped to make sure that the results were representative of the 'universal features of adult human mental life.'"[17] Viewing the subject as having a generalized mind meant that experimenter and subject could switch roles between trials without affecting the format of the experiments.

In sum, as psychologist Edward Titchener explained in 1912, it was not that "the subject should be hooked up to machines"; it was that the subject had "virtually become the machine, capable of automatic introspection."[18] With training the subject could register the moment at which he had recognized a stimulus and thus reveal the reaction time between the appearance of the stimulus and the mind's psychological, introspective recognition of the stimulus.

### Torres Straits Islands: The "Generalized Mind"

THESE ISSUES WERE also apparent in the Cambridge Anthropological Expedition to the Torres Straits Islands in 1898, an event in which understandings and practices sympathetic to Wundt's introspection were continued, and an event that some have considered to be foundational for fieldwork in social anthropology. Since the expedition's scientists assumed that the social and natural environment determined the way the mind perceived the world, they also assumed that after they had immersed themselves in the daily life of villagers on the islands, they could serve as appropriate experimental subjects comparable to the native inhabitants.[19] Their introspective reports of the time they took to react to a stimulus were measured and compared to the reports of native Torres Straits Islanders in hand-drawn spread sheets, later published in their reports—another kind of "table" of course.

In the Cambridge Expedition, as in European labs of that time, experimenters and subjects could trade places. In a photograph of W. H. R. Rivers with the color wheel, Rivers and his companion (his name is Tom) are on the same side of the table: Rivers is not studying Tom but showing him how to use the color wheel.[20]

The assumption that informed Rivers's work, according to Henrika Kuklick, was that "a resident of the Torres Straits Islands was no different from any of Rivers's experimental subjects—including Rivers himself." His notion of a generalized mind (extended to these islands) entailed that the context in which such minds were trained determined their specific characteristics! Many do not realize that Rivers trained himself to participate with the "minds" of Torres Islanders: he imagined he could immerse himself in their lives and "faithfully follow their way of life." He "would serve as a one-person research instrument because he would literally think and feel as they did." I am suggesting that there is *resonance* between these ideas and the ideas behind Wundt's laboratory training, aimed to make subjects comparable through experience of and training in the same regimen. In the Torres Straits the regimen entailed immersion in the environment and social life of the islanders.[21]

This led them to some important discoveries: at the time of the Torres Straits expedition, the psychologists on the team (Rivers and C. S. Myers) were haunted by the widely accepted evolutionary theories of Herbert Spencer that "'primitives' surpassed 'civilised' people in psychophysical performance because more energy remained devoted to this level in the former instead of being diverted to 'higher functions,' a central tenet of late Victorian 'scientific racism.'"[22] The expedition reports refer to this theory, and their disproof of it. Smell was shown to be less acute among the Islanders, and so was hearing. Their less acute hearing was put down to aspects of the social environment, namely the amount of time they spent diving for pearls—an activity they had been forced into by European (and increasingly Japanese) traders.[23]

## C. S. Myers

TRAINING IN THE Wundtian sense continued in Cambridge after the expedition. Myers's studies in the Torres Straits Islands and later in the Cambridge Laboratory of Experimental Psychology focused on aural perception in music and rhythm.[24] He founded the psychological laboratory at Cambridge in 1912, taught experimental psychology, and authored a two-volume textbook on the subject. He was interested not just in recording music, measuring its intervals, and measuring reaction times in various sensory modalities but specifically in the subjective components of sensory experience. For example, using a Wundtian apparatus in Cambridge he could present sounds with various intervals in between; the subject would try to replicate the intervals, and these would be recorded on the smoked surface of a revolving drum: "The subject should carefully record the results of introspective analysis."[25]

Throughout his career, well into the 1930s, Myers stressed that the aesthetic aspects of music and rhythm had to be understood comparatively in different cultures.[26] In "The Ethnological Study of Music" he summarizes, "Thus it comes about that many examples of primitive music are incomprehensible to us. . . . Our attention is continuously distracted, now by the strange features and changes of rhythm, now by the extraordinary colouring of strange instruments, now by the unwonted progression and character of intervals. . . . We have first to disregard our well-trained feelings towards consonances and dissonances. . . . Thus incomprehensibility will gradually give place to meaning, and dislike to some interesting emotion."[27]

Whereas in Wundt's lab and the Torres Straits expedition, training was instilled to make subjects comparable, in Myers's lab previous training was extinguished to make music comprehensible. But in both cases introspection was central.

## Edward Titchener

TRAINING IN THE Wundtian sense was vigorously championed in the United States by Edward Titchener, who wrote the textbooks and manuals used in American labs for decades in the early twentieth century and represented the apotheosis of introspection. He wrote that the validity of the experiment rested largely with the subject (whom he called the Observer). (He explicitly directed the subject to observe his internal mental states.) If the subject was aware of relaxing his attention, of succumbing to intruding ideas, or of being affected by disturbances in the room, he must "note the departure from the norm, and throw out the [data that result]." The experimenter was to "sacrifice unquestioningly" such results, whether or not they seemed good to him. Under these assumptions it made sense that each portion of an experiment should begin with a "good number of practice series, the results of which are not counted," to give the subject time to "warm up" or "get into swing."[28] What they called "practice effects" were a good thing, then.

As is well known, not long after Titchener wrote so emphatically about introspective methods, behaviorism in American psychology rang the death knell for any such reliance on the subject's introspective experience. Today in my fieldwork any mention I might make of the role of subjectivity or introspection in an experimental model is roundly rejected. Individual differences are washed out by the normative variation across subjects, and subjects simply cannot know anything about the unconscious processes taking place in their cognition.

## Summary

MY PARTIAL SOLUTION to the puzzle I began with is that practices from a previous epistemic regime (when conscious introspection was central) persisted into a new one (when unconscious responses are central). Practice was required for Wundt's students and the Cambridge anthropologists in order to make the introspective experience of subjects comparable—but is submerged in today's experiments alongside knowledge that practice of any kind dramatically affects the outcome of tests and alongside insistence that interventions of the researcher in the process of capturing nature are undesirable.

Daston and Galison's two epistemic regimes need to be slightly revised to fit experimental psychology as follows. In the earlier era, when introspection (conscious experience) mattered, subjects had to be trained so their reactions would be comparable. Wundt's and the Cambridge Expedition's subjects (both European and otherwise) needed to have "trained judgment." Meanwhile the structure of the experiment itself would keep researchers from intervening in the capture of nature. In the subsequent era, after introspection was ruled out of psychological experiments, you would think (logically) that subjects would no longer need "trained judgment" since only unconscious experience was being measured. In neither the earlier nor the later era would it be considered proper for the researcher to intervene in the structure of an experiment itself. Paradoxically, in the later era, including the present, subjects are still trained, risking the possibility that experimenters *are* intervening in the experiments because of the known effects of training on testing results.

My goal in thinking about this is certainly not to (as researchers fear) "make them look bad." Our own fields are surely full of such paradoxes, and they are valuable wherever they occur as a wedge into anthropological understanding. The paradox here is that human perception can be treated as if it were *not* subjective.

It is as if the armory of techniques to hold subjects stable and thus comparable (bite bar, fixation point, tables, etc.) is insufficient. Perhaps just as the graphic form of the table places the variables in the experiment on a flat plane so they can be compared in the same artificial space, so practice trials, taking place over time, bring the subjects up to a common base so their reaction times can be compared during the experiment. This extra measure of control increases the chances of achieving the goal experimenters seek: a statistically significant objective "effect."

The psychologist Martin Orne had it right when criticizing behaviorism in the 1960s. He argued that, claims among his contemporaries to the contrary, the subject is not entirely passive in the "experimental setting."[29] Researchers unintentionally "demand" certain characteristics from their subjects, the way I felt the "demand" to be a good student during my training. The narrow concept of the experiment, with its controlled variables and techniques of stabilizing subjects, takes place in a larger setting: one could say the experiment is preceded by a foyer or entrance hall in which the subject is trained to be a good subject.

In the first fieldwork incident I described, I was told that my subjective experience had been left behind at the door, but it would be more accurate to say that inside the door was an anteroom where training would take place and some elements of subjective experience would linger. These lingering elements of subjectivity are what Betty Bayer describes as the phantoms that have apparently been eliminated from psychological research but actually continue to haunt it.[30] These phantoms are what make it seem necessary for the researcher to intervene in the "capture of nature" and harness training to the same kind of ends as the fixation point, the table, and the experimental model.

# Endnotes

1. I am grateful to David Bates and the participants in the conference Plasticity and Pathology: History and Theory of Neural Subjects at the University of California, Berkeley, where this paper first took shape. Parts of the paper about the Cambridge Expedition to the Torres Straits were published, in another form, as Emily Martin, "The Potentiality of Ethnography and the Limits of Affect Theory," *Current Anthropology* 54, supplement 7 (2013).

2. Lorraine Daston and Peter Galison, *Objectivity* (New York: Zone, 2007), 21.

3. B. Kantowitz, H. Roediger, and D. Elmes, *Experimental Psychology* (Stamford, CT: Cengage Learning, 2014), 244. David Martin recommends "practice trials" to "minimize warm-up effects," which he defines as "fast improvement" before "general readiness," D. Martin, *Doing Psychology Experiments* (Stamford, CT: Cengage Learning, 2007), 31,152. In G. M. Breakwell, J. A. Smith, and D. B. Wright, *Research Methods in Psychology* (Los Angeles: Sage, 2012), 177–78, "general tips" are provided that advise experimenters to use practice trials to make sure the performance levels off before you capture the "variability intrinsic to performance." Breakwell et al. also state that it will take a lot of practice trials before "performance stabilizes."

4. J. Carson, "Minding Matter/Mattering Mind: Knowledge and the Subject in Nineteenth-Century Psychology," *Studies in History and Philosophy of Science Part C: Studies in History and Philosophy of Biological and Biomedical Sciences* 30, no. 3 (1999): 351.

5. Kurt Danziger, *Constructing the Subject: Historical Origins of Psychological Research* (Cambridge: Cambridge University Press, 1990), 183.

6. Daston and Galison, *Objectivity*, 378.

7. H. von Helmholtz, *Handbuch der physiologischen Optik: Mit 213 in den Text eingedruckten Holzschnitten und 11 Tafeln* (Leipzig: Voss, 1866), 42.

8. Bruno Latour, "Circulating Reference: Sampling the Soil in the Amazon Forest," in *Pandora's Hope: Essays on the Reality of Science Studies* (Cambridge, MA: Harvard University Press, 1999), 38.

9. Alfred Gell, "Vogel's Net," *Journal of Material Culture* 1, no. 1 (1996): 15–38.

10. Latour, "Circulating Reference," 30.

11. Deborah J. Coon, "Standardizing the Subject: Experimental Psychologists, Introspection, and the Quest for a Technoscientific Ideal," *Technology and Culture* 34, no. 4 (1993): 770.

12. Ruth Benschop and Douwe Draaisma, "In Pursuit of Precision: The Calibration of Minds and Machines in Late Nineteenth-Century Psychology," *Annals of Science* 57, no. 1 (2000): 20.

13   Ibid., 22.

14   Coon, "Standardizing the Subject," 775.

15   Benschop and Draaisma, "In Pursuit of Precision," 18–19.

16   James M. K. Cattell and Michael M. Sokal, *An Education in Psychology: James McKeen Cattell's Journal and Letters from Germany and England, 1880–1888* (Cambridge, MA: MIT Press, 1981), 89.

17   Benschop and Draaisma, "In Pursuit of Precision," 58–59.

18   Qtd. in Coon, "Standardizing the Subject," 776.

19   A. Herle and S. Rouse, "Introduction: Cambridge and the Torres Straits," in *Cambridge and the Torres Strait: Centenary Essays on the 1898 Anthropological Expedition*, ed. Anita Herle and Sandra Rouse (Cambridge: Cambridge University Press, 1998); Henrika Kuklick, "Fieldworkers and Physiologists," in Herle and Rouse, *Cambridge and the Torres Strait*; Henrika Kuklick, "Personal Equations: Reflections on the History of Fieldwork, with Special Reference to Sociocultural Anthropology," *Isis* 102 (March 2011): 1–33; Graham Richards, "Getting a Result: The Expedition's Psychological Research 1898–1913," in Herle and Rouse, *Cambridge and the Torres Strait*.

20   Richards, "Getting a Result," 143.

21   Kuklick, "Fieldworkers and Physiologists," 174, 175.

22   Richards, "Getting a Result," 137.

23   Alfred C. Haddon et al., *Reports of the Cambridge Anthropological Expedition to Torres Straits*, vol. 2: *Physiology and Psychology* (Cambridge, UK: University Press, 1901), 14, 42; A. C. Haddon, *Head-Hunters: Black, White, and Brown* (London: Methuen, 1901), 121–22.

24   Barbara W. Freire-Marreco and John Linton Myres, *Notes and Queries on Anthropology* (London: Royal Anthropological Institute, 1912).

25   C. S. Myers, *A Text-Book of Experimental Psychology with Laboratory Exercises*, Part II: *Laboratory Exercises* (Cambridge: Cambridge University Press, 1911), 97.

26   Charles S. Myers, *In the Realm of Mind* (Cambridge: Cambridge University Press, 1937), 63.

27   Charles S. Myers, "The Ethnological Study of Music," in *Anthropological Essays Presented to Edward Burnett Tylor in Honour of His 75th Birthday, Oct. 2, 1907*, ed. Edward Burnett Taylor, Henry Balfour, and Barbara W. Freire-Marreco (Oxford: Clarendon, 1907), 249.

28   E. B. Titchener, *Experimental Psychology: Students' Manual*. Part 2: *Instructor's Manual* (London: Macmillan, 1910), 4, 150, 218, 360.

29 Martin Orne, "Demand Characteristics and the Concept of Quasi-controls," in *Artifact in Behavioral Research*, ed. R. Rosenthal and R. L. Rosnow (New York: Academic Press, 1969).

30 Betty M. Bayer, "Between Apparatuses and Apparitions: Phantoms of the Laboratory," in *Reconstructing the Psychological Subject*, ed. Betty M. Bayer and John Shotter (London: Sage, 1998), 187.

*Catherine Malabou*

# 2 "You Are (Not) Your Synapses": Toward a Critical Approach to Neuroscience

TWO RELATIVELY RECENT and perfectly simultaneous intellectual encounters happened to be decisive for my philosophical trajectory, changing its course and making any attempt at going backward impossible: my encounter with current neuroscience, on the one hand, and my encounter with the thought of Michel Foucault, on the other. I had read Foucault before, of course, but my knowledge and experience of his philosophy had remained shallow until I started exploring neuroscientific literature and asked myself whether it was possible to constitute the brain as an object for continental, that is nonanalytical or noncognitivist, philosophy.

I would like to show here that the elaboration of the encounter between these two encounters might determine a new starting point for contemporary thought.

This in-between space remains improbable and difficult both to situate and clear though, to the extent that it appears at first sight more a battlefield or a war front than an exchange or a discussion platform.

Let's start with the war, then, and let's gradually show how the conflict can be pacified. Isn't this, after all, the task of critique?

Critique is precisely Foucault's topic in a short but fundamental text from 1978, "What Is Critique?," revised in 1984 and renamed "What Is Enlightenment?": "In November 1784 a German periodical, *Berlinische Monatschrift*, published a response to the question: '*Was ist Aufklärung* [What is Enlightenment]?' And the respondent was Kant." "Let us imagine," Foucault goes on, "that the *Berlinische Monatschrift* still exists and that it is asking its readers the question: . . . what, then is this event that is called the *Aufklärung* and that has determined, at least in part, what we are, what we think, and what we do today?"[1]

Let us address this issue again in 2014: What is this event that might be called *Aufklärung*, Enlightenment, today?

Let us then try to imagine the most improbable of all answers, the worst answer, from a continental philosopher's point of view. I borrow this terrible answer from the neurobiologist Joseph LeDoux, who, in his book *Synaptic Self*, writes, "You are your synapses." To the question "What is Enlightenment today?" the answer would then be: *Aufklärung* is a synaptic process. "You are your synapses. They are who you are," LeDoux goes on. "The key to human reason is to be found in the microscopic spaces between two nerve cells."[2]

What? How might *Aufklärung*, thought, reason, freedom be rooted in the brain? How might critique—which Kant presents, as Foucault reminds us, as *Aufklärung* itself—how might critique emerge from our synapses?

"You are your synapses": Isn't this the most reductionist, obscure, and obscurantist of all answers? And isn't that the reason continental philosophers so often reject neurobiology as a possible interlocutor?

Let us now imagine a dialogue between a continental philosopher and a cognitivist philosopher—a cognitician—as they discuss this same statement: "You are your synapses." Foucault will stand for the continental philosopher; Thomas Metzinger, the famous German cognitician, a great reader of LeDoux, will stand on the other side. The dialogue will examine every term of the statement: "you," "are," "your," "synapses."

## You

*AUFKLÄRUNG* COINCIDES, AS Foucault demonstrates, with the constitution of the self as an autonomous subject. To state that subjectivity might depend upon empirical and biological data amounts to arguing that the self is, on the contrary, heteronomous, and this would represent an escape from the principles of rationality. The "you," the subject, the identity, the self, that critique tries to bring to light and to emancipate is and can only be the free subject.

The situation is even worse than you think, the cognitician answers. There is no such thing as a subject. "You are your synapses" simply means that there is no "you." Such is the main thesis that Metzinger develops in his book called, very eloquently, *Being No One*.[3] You are your synapses means you are no one, nobody. No subject. Consequently, also, autonomy is never yours.

Metzinger declares that "nobody ever *was* or *had* a self. . . . The phenomenal self is not a thing, but a process."[4]

All right, the continental philosopher would reply, but Kant emphasized, centuries before you, the nonsubstantial character of the subject.

If this is the case, then, Metzinger would answer in his turn, how can you affirm that autonomy, from a critical point of view, stems from "three broad areas: relations of control over things, relations of action upon others, relations with oneself"?[5] If we take for granted, after Kant, that the self is not substantial, what are "relations with oneself" supposed to mean? You still seem to believe that subjectivity is self-related, auto-affected. That subjectivity, in other words, coincides with consciousness.

Being no one, Metzinger pursues, means that consciousness, consequently also the idea of a relation with oneself, is an illusion. Such an illusion is created by specific neural processes, included in a structure that Metzinger calls neural transparency, or the transparent self-model. Commenting upon *Being No One*, Slavoj Žižek writes, "The self is its own appearance, [Metzinger] writes, since it is a model which cannot perceive itself as a model, and thus exists only insofar as it does not perceive itself as a model."[6] Metzinger declares, "What in philosophy of mind is called the 'phenomenal self' and what in scientific or folk-psychological contexts frequently is simply referred to as 'the self' is the content of a phenomenally transparent self model. The subjective experience of *being someone* emerges if a conscious information-processing system operates under a transparent self-model. . . . It is transparent: you look right through it. You don't see it. But you see *with* it. . . . You constantly *confuse* yourself with the content of the self-model currently activated by your brain."[7]

The first-person perspective is not an origin but the result of a series of multiple progressive biological processes. These processes are paradoxically doomed to disappear from the realm of consciousness. Some other processes in our brain "swallow" and "erase" all of the processing stages that were necessary for the construction of consciousness and the first-person perspective. Such an erasure "is activated in such a fast and reliable way as to make any earlier processing stages inaccessible to introspection." The experience of the immediacy of consciousness thus emerges as a "temporal fiction" that proceeds from the impossibility of consciousness to have access to its own biological past.[8] Neural processes produce this effect of speed, which renders introspection impossible. We cannot turn back on ourselves. If we were able to do so and access the neural stages of consciousness's formation, we would get caught in a maze, which would immediately interrupt consciousness. The brain then has to find an efficient way to break the reflexive labyrinth, to make information available without engaging the system in endless internal loops.[9] Neural transparency and the invisibility of self-modeling thus coincide with the nonexistence of subjectivity: the erasure of presubjective processes creates a window through which we see *while never seeing the glass of the window itself.* In this sense "transparency is a special form of darkness."[10]

Transparency is what renders critique impossible as a reflexive kind of process. By definition, transparency cannot see itself. Which also means that consciousness cannot accurately perceive its own relation to the brain. As Žižek again notes in *The Parallax View,* "It is in the nature of consciousness that it misperceives the gap which separates it from raw nature. That is to say, when I experience myself directly as a self, I by definition enact an epistemically illegitimate short circuit, misperceiving a representational phenomenon for reality. As Lacan put it, with regards to the ego, every cognition is misrecognition."[11]

The neuroscientific approach to subjectivity challenges critique to the extent that it raises the paradox of an entity that exists only insofar as it remains unknown, the paradox of an asubjective origin of subjectivity.

## Are

NOW, RESPONDS OUR continental philosopher, what about Being, what about the ontological status of the formula "You *are* your synapses"? How could a biological miracle put an end to the meaning of Being? What about ontology, then?

The fact that critique, for Kant, cannot by any means be identified with any dogmatic ontology does not prevent Foucault from defining contemporary *Aufklärung* as a "historical ontology of ourselves" or a "critical ontology of ourselves." Such an ontology, according to Foucault, has to provide answers to the following issues: "How are we constituted as subjects of our own knowledge? How are we constituted as subjects who exercise or submit to power relations?

How are we constituted as moral subjects of our own actions?"[12] In other terms, how do we become what we are?

The historico-critical ontology of ourselves has to be considered not, certainly, as a theory, a doctrine, nor even a permanent body of knowledge but as a practical and political reelaboration of the concept of freedom. Ontology in that sense has to be understood as an art of existence, of becoming, not so much as a doctrine of being understood as a genealogy of ways of being. It "has to be conceived as an attitude, an *ethos*, a philosophical life in which the critique of what we are is at one and the same time . . . a work carried out by ourselves upon ourselves as free beings."[13]

We can of course wonder what freedom and free will become within the strictly cognitive framework. How could there be any space for freedom in neurophilosophy if all our decisions are predetermined by transparent, that is obscure, objective brain processes? Biological determinism eliminates the very possibility of conferring an ontological status to our existence and mental acts. If there is something like an ontology, Metzinger argues, it can only be what he calls a "motor" or "functional ontology,"[14] a new kind of mechanism, which substitutes speed and movement for Being. Ontology could never be originary or quick enough to recapitulate biology. Ontology in that sense would always come a posteriori. Being no one also means that no one is, that there is no such thing as being.

## Your

WHAT ABOUT *YOUR* synapses, then? Is there any sense in affirming that *my* synapses belong to *me*? Metzinger stresses the impossibility of conferring a determined meaning to an expression like "my brain": "There is no way the subject, from the 'inside,' can become aware of his own neurons, from the 'inside.' They can be known only objectively, from the 'outside.' There is no inner eye watching the brain itself, perceiving neurons and glia. The brain is 'transparent' from the standpoint of the subject."[15]

The only possible access to this transparency is when it is impaired, that is, when brain damage occurs. That is why neurobiologists often characterize cerebral lesions as a "method" to explore the brain from inside. But of course when neural transparency or the impossibility of turning back on oneself are impaired, what happens is not the emergence of a truth but, on the contrary, the emergence of a total indifference to truth. The "survivors of neurological disease," as Antonio Damasio calls them in *The Feeling of What Happens*, sometimes lead a life whose temporality and structure are almost totally destroyed.[16] Breaking the glass does not reveal any critical secret. If we can say "I grasp with my hands" and "I see with my eyes," we cannot really say "I think with my brain," because the expression *my brain* cannot be associated with any experience. Nobody can feel his or her own brain; the brain does not belong to the body proper. The brain is never here, but always there, nowhere—it has no subjective site.

But does subjectivity in general have any subjective site, asks the continental philosopher? We have known since Kant, as he states in §24 of the *Critique of Pure Reason*, that the subject is both a logical form (the form of the "I think"), with no sensuous content, and the empirical form of the subject's intuition, the way the subject "sees," "feels," or "intuits" herself.

It means that the subject can represent itself only as it appears empirically to itself. Second, the "I think," or apperception, as soon as it takes itself as an object, loses its formal logical determination to become an intuited object, that is, an object of its own inner sense, which affects it. The subject can represent itself only as affected—that is, also altered—by itself. The self has access to itself through its own otherness or alterity. When I say "myself," when I feel or think of "my" identity, I express only the result of this nonessential and altering self-representation.

Still, the cognitician would reply, Kant calls this self-representation an "auto-affection." There still remains something like an

"I" or a "me" in this process, and we just saw that any intuition of an "auto," of a self, is in reality delusive.

## Synapses

LET'S THEN EXAMINE the last term, *synapses.*

Surprisingly this last word allows us to gradually reverse all previous arguments and to situate the confrontation between philosophy, critique, and neurobiology quite differently. The term *synapse* was fashioned in 1897 by the English physiologist Charles Sherrington from the Greek word *synaptein,* meaning "to fasten together." *Synaptein* is combined from the Greek *syn-,* "together," and *haptein,* "to fasten or bind." Synapses are the spaces between brain cells. How does a void, a space, a cleft, come to be called a synapse if this term etymologically refers to a kind of synthesis and gathering? How can such intervals be considered the channels of communication between cells that make possible all brain functions, including perception, memory, emotion, and thinking? How can they be defined as the main sites of storage of information, the information that is encoded by our genes and also by our experiences—our memories?

Historically it was generally thought that the role of the synapse was to simply transfer information between one neuron and another. We now know, since the remarkable work of Santiago Ramón y Cajal in Spain and Donald's Hebb book *The Organization of Behavior,* published in 1949 in New York, that the efficacy of synaptic transmission is not constant; it varies depending upon the frequency of stimulation. The modulation of synaptic efficacy provokes modifications of neural connections in volume, size, and shape. When the connections are frequently stimulated, by playing piano regularly, for example, they increase in size and volume (this phenomenon is known as long-term potentialization). In contrast when certain connections are no longer activated, they decrease, and scientists talk about synaptic long-term depression.

This phenomenon of modulation is called synaptic or neural plasticity. Neuronal circuits are capable of self-organization. LeDoux writes, "The efficacy of the synapses varies with respect to the flux of information traversing them: during infancy and throughout life, each one of us is subject to a unique configuration of influences from our external surroundings which resonates in the form and the functioning of the brain's network."[17] The fact that the synapses have their efficacy reinforced or weakened as a function of experience thus allows us to assert that, even if all human brains resemble each other with respect to their gross anatomy, no two brains are identical with respect to their history, as the phenomena of learning and memory show directly.

The modification of neural connections due to synaptic modulation is called neural plasticity. Plasticity acts like a sculptor, and we can speak of a plastic art of and in the brain.

The term *plasticity*, the meanings of which have been central to my work for a long time now, is what allows us to mediate the conflict between cognitivism and philosophy.

We see that, paradoxically, if neural transparency causes the illusion of subjectivity and reveals that we are no one, neural plasticity, which is intimately linked with neural transparency, on the contrary situates the brain as the core of our individual experiences and identity. We may be no one, but this impersonality is plastic, which means that this absence of subjectivity is paradoxically malleable, fashionable, so that each of us is no one in his or her own way.

How can we understand such a paradox, however? How is it possible to bridge the gap between transparency and plasticity?

Returning to Foucault's text allows us to give an answer. It is now time to approach "What Is Enlightenment?" differently.

How can subjectivity be said to be nonsubstantial and plastic at the same time? Foucault answers: because subjectivity consists only in its own formation and transformation. A critical approach to subjectivity reveals the priority of fashioning over being. If subjectivity does not exist, because it doesn't exist, we have to invent it.

Inventing oneself: this idea orients Foucault's whole reading of Kant's *Was ist Aufklärung?*

According to Foucault, in this text, Kant, for the first time in history, opens the possibility for European philosophy to reflect on its own present. The "present" then becomes, for the first time, a philosophical category in the form of the following question: What difference does today introduce with respect to yesterday? According to Foucault, "[It] is in the reflection on 'today' as difference in history and as motive for a particular philosophical task that the novelty of this text appears to me to lie. And, by looking at it in this way, it seems to me we may recognize a point of departure: the outline of what one might call the attitude of modernity."[18]

If there is such a thing as a present for philosophy, this present cannot be a deduction; it must be an invention: "Modern man . . . is not the man who goes off to discover himself, his secrets and his hidden truth; he is the man who tries to invent himself. Modernity does not 'liberate man in his own being'; it compels him to face the task of producing himself." Modernity thus implies "a change that man will bring upon himself."[19]

Further, "Kant, in fact, describes Enlightenment as the moment when humanity is going to put its own reason to use, without subjecting itself to any authority; now, it is precisely at this moment that the critique is necessary, since its role is that of defining the conditions under which the use of reason is legitimate in order to determine what can be known, what must be done, and what may be hoped."[20]

If humanity has to make use of its own reason freely for the first time in history, this implies that this use has to be invented. Free reason has to invent itself. Because it does not exist as a substance or a pregiven structure, it consists only in its own plastic self-constitution. It receives the very form that it gives to itself.

In that sense the present is a synapse, a temporal gap that has to be formed and filled with a new shape, and this new shape is the form of our reason, that is, of ourselves.

Foucault's concepts of technologies of the self, self-stylization, self-fashioning, all refer to this idea, according to which subjectivity never precedes its own invention. The critical ontology of ourselves, previously evoked, is then to be understood as this very gesture of creation. What, given the contemporary order of being, can I be?

If it is true that the self fashions itself in terms of existing norms, always comes to inhabit and incorporate these norms, including the biological ones—what Foucault calls subjectivation (*assujettissement*)—we have to understand that the norm is not external to the principle by which the self is formed. As Judith Butler puts it in her essay "What Is Critique? An Essay on Foucault's Virtue":

> Critical practice turns out to entail self-transformation in relation to a rule of knowledge or a rule of conduct. . . . Foucault asks what criteria delimit the sorts of reasons that can come to bear on the question of obedience. He will be particularly interested in the problem of how that delimited field forms the subject and how, in turn, a subject comes to form and reform those reasons. This capacity to form reasons will be importantly linked to the self-transformative relation. . . . To be critical of an authority . . . requires a critical practice that has self-transformation at its core."[21]

If subjectivation supposes a submission to a form—be it a rule, a law, or any kind of cognitive pattern or template—it also supposes what Foucault calls a desubjugation (*désasujettissement*) by which this form is re-created and remodeled.

Isn't desubjugation close to the erasure of subjectivity evoked a moment ago by Metzinger? At the end of his book Metzinger admits that self-modeling is not simply the result of an adaptation to a biologically given form. More generally, neural plasticity is certainly not subsumed within a naturally and culturally given form. One fashions oneself only by resisting given norms. Each brain's style, or neural singularity, due to synaptic plasticity, is the result of such a resistance. Here is the essential difference between plasticity and flexibility. To be flexible is to receive a form or impression, to be able to form oneself, to take the form, not to give it. To be docile, not to resist. No scientist would ever speak of neural flexibility. The scientific concept is neural plasticity, which integrates creativity as an objective dimension of the brain.

Metzinger acknowledges such a creativity. After he discusses the absence of self and auto-affection, he surprisingly writes, "No one ever was or had a self. As soon as the basic point has been grasped . . . a new dimension opens. At least in principle, one can wake up from one's own biological history. One can grow up, define one's own goals, and become autonomous. And one can start talking back to Mother Nature, elevating her self-conversation to a new level."[22]

I cannot resist the pleasure of quoting Žižek commenting on that very passage: "Surprisingly, we thus encounter, at the very high point of a naturalistic reductionism of human subjectivity, a triumphant return of the Enlightenment theme of a mature autonomous . . . what? Certainly not self but artificial subjectivity."[23]

Plasticity would then perhaps constitute the link between self-invention and artificial subjectivity. Between critical and neurobiological Enlightenment. In his way Metzinger would then have answered the question "What is *Aufklärung*?"

The last and certainly the most important and looming issue that remains to be addressed as a concluding development is that of the role of the transcendental.

Wouldn't our continental philosopher ask at that point: How can there be any critique without transcendentality, that is without a priori, and certainly not biological, laws and limits of reason for this self-fashioning?

In fact no, this is not what he would ask. Here is what Foucault asks: What if the transcendental itself was plastic? What if the a priori itself was malleable, that is, in a way, historical, and consequently also contingent?

This question haunts the most beautiful passages of Foucault's text: "If the Kantian question was that of knowing what limits knowledge [*connaissance*] must renounce exceeding, it seems to me that the critical question today must be turned back into a positive one: In what is given to us as universal, necessary, obligatory, what place is occupied by whatever is singular, contingent, and the product of arbitrary constraints?"[24]

These issues are of course linked with the problem of self-transformability: "[The new] criticism is no longer going to be practiced in the search for formal structures with universal value but, rather, as a historical investigation into the events that have led us to constitute ourselves and to recognize ourselves as subjects of what we are doing, thinking, saying. In that sense, this criticism is not transcendental."[25]

It would then be possible to transform the critique conducted in the form of a necessary limitation into a practical critique that would take the form of a "crossing-over."[26] It would then become possible to transgress the transcendental, which would amount to redrawing the borders between the transcendental and experience. Foucault affirms that "this historico-critical attitude must also be an experimental one."[27]

To experiment with the transcendental: such would be the imperative for philosophy today.

Who said that? "I did," says Metzinger. "No one," says Foucault. Who is speaking then?

Nobody knows; the speaker remains to be invented.

PHILOSOPHY REMAINS TO be invented. Refusing to draw strict, rigid, or dogmatic limits between the brain and consciousness, between biology and history, between natural and symbolic life requires philosophy to deconstruct the very concept of limit and work in and on the unstable space of this deconstruction; it requires that we play with the plastic material such a space provides.

What is *Aufklärung* today? *Aufklärung* is what allows us to experiment with the transcendental, to put the plasticity of reason to the test, from both the continental and the neurobiological point of view.

Continental philosophy provides us with the tools, the concepts, and the theoretical framework that allow us to bring to light and examine the implications of the plasticity of our—artificial and free, artificial but free—selves. Conversely a neurobiological approach to philosophy offers us the possibility of eliminating our symbolic inhibitions that make us regard certain limits as precisely undeconstructible. In the end the philosopher has to invent a new form of identity that cannot belong any longer to the continental or cognitivist traditions but that has to transgress this division itself.

Instead of "You are your synapses," I might have chosen another formula, which has become a neuroscientific motto: "Cells that fire together wire together." This playful sentence summarizes the theory of synaptic plasticity. This means that when particular nerve cells are activated at the same time by a stimulus such as a sound or smell, they strengthen their physical ties with each other. These changes in the interconnections among nerve cells in the brain allow for constant neural plasticity.

There is a way continental philosophy and neuroscience, critical and empirical thought, the biological and the symbolic can fire and wire together. The encounter between these two fields opens one of the most exciting theoretical, practical, and political adventures of our time. It allows a genuine dialogue that simultaneously respects the autonomy of each field and redraws their mutual limits. How can we now come to think of a neuroplasticity of the humanities that would bring some plasticity to the humanities as well as some critical theory to neurobiology?

## Endnotes

1. Michel Foucault, "What Is Enlightenment?," in *The Essential Foucault,* ed. Paul Rabinow and Nikolas Rose (New York: New Press, 2003), 43.

2. Joseph LeDoux, *Synaptic Self: How Our Brains Become Who We Are* (London: Macmillan, 2003), 324.

3. Thomas Metzinger, *Being No One: The Self-Model Theory of Subjectivity* (Cambridge, MA: MIT Press, 2003).

4. Ibid., 1.

5. Foucault, "What Is Enlightenment?," 55.

6. Slavoj Žižek, *The Parallax View* (Boston : MIT Press, 2006), 214.

7. Metzinger, *Being No One,* 1.

8. Ibid., 98, 57.

9. Ibid. Cf. 337 et seq.

10. Ibid., 169.

11. Žižek, *The Parallax View,* 215.

12. Foucault, "What Is Enlightenment?," 53, 56.

13. Ibid., 49.

14. Metzinger, *Being No One,* 317 et seq.

15. Todd Feinberg, quoted in Metzinger, *Being No One,* 177.

16. Antonio Damasio, *The Feeling of What Happens: Body and Emotions in the Making of Consciousness* (New York : Harcourt, Brace, 1999), 313.

17. LeDoux, *Synaptic Self,* 177.

18. Foucault, "What Is Enlightenment?," 48.

19. Ibid., 50.

20. Ibid., 47.

21. Judith Butler, "What Is Critique? An Essay on Foucault's Virtue," in *The Political: Readings in Modern Continental Philosophy,* ed. David Ingram (London: Basil Blackwell, 2002), quoted here from the Instituto Europeo para Políticas Culturales Progresivas electronic version.

22. Metzinger, *Being No One,* 634.

23. Žižek, *The Parallax View,* 221.

24. Foucault, "What Is Enlightenment?," 53.

25. Ibid.

26. Ibid.

27. Ibid., 54.

*Cathy Gere*

# 3 Plasticity, Pathology, and Pleasure in Cold War America

IN 1970 TULANE University neuroscientist Robert Heath attempted to reorient a young gay man's sexuality by means of direct electrical stimulation of his neural "pleasure center." The experiment combined a heteronormative view of healthy sexual functioning with a commitment to radical neural plasticity, both grounded in an evolutionary theory of reward and reinforcement. According to this theory, pain and pleasure are evolved mechanisms that guide sentient beings toward adaptive and away from maladaptive stimuli. Because homosexuality is nonprocreative, and therefore self-evidently maladaptive, it must be a "disorder of hedonic regulation," in which pleasure and desire are directed to unhealthy objects. Pleasure and pain were also the targets for treatment: Heath believed he could redirect maladaptive behavior by producing new appetites and aversions with electrical stimulation of the neural reward centers.

Arguing that reward and punishment were key terms in cold war debates about neural plasticity, this paper examines some key episodes in hedonist psychology, culminating in Heath's 1973 publication, "Septal Stimulation for the Initiation of Heterosexual Behavior in a Homosexual Male." Intercut with this account is the story of hedonist psychology's opponents, defenders of an ideal of human autonomy, drawn from a variety of social movements and scholarly schools, including some of the founders of cognitive science. These scientists championed a view of the innate, genetically determined structure of the human brain that accounted for our species-specific attributes, including our moral freedom.

Both sides of the debate drew from evolutionary theory, but in the context of cold war anxieties about brainwashing and mind-control techniques, it was the hedonists' extreme commitment to behavioral plasticity that made them vulnerable to accusations of totalitarianism. Between World War II and Watergate, in other words, the ideological battle lines in psychology were the inverse of our contemporary fascination with neuroplasticity as "the biology of freedom." The paper concludes with an open question about what we might want to take away from this historical reversal.

## Operant Conditioning

THEORIES ABOUT THE behavioral effects of pain and pleasure are as old as the discipline of neurology, but it was not until the turn of the twentieth century that hedonist psychology was systematized into an experimental practice. In 1898 the American psychologist Edward Thorndike published *Animal Intelligence*, laying out his theory of reward and learning. He recounted how he had placed cats into boxes of his own construction, with doors that opened when pressure was put on a simple pulley-and-latch mechanism, and dangled a piece of fish outside the door as a reward. The cats would start out clawing and scratching at the walls of the box at random. If they accidentally pulled the latch, they found themselves suddenly released from confinement and able to reach the food. By timing how long each cat took to free itself from the puzzle box in a series of repeated trials, and showing how gradual was the "learning curve" in each case, Thorndike argued that the animals mastered the task not by reasoning or imitation but by a process he called the "Law of Effect," in which any action that resulted in a pleasurable outcome—such as reaching the food and eating it—was liable to be repeated.

In 1913 Thorndike's reductive approach got a name, and an even more radical set of philosophical commitments, when the Johns Hopkins professor of psychology John Watson published a polemical paper, "Psychology as the Behaviorist Views It," exhorting psychologists to recognize "no dividing line between man and brute."[1] Watson was inspired by the Russian scientists Ivan Sechenov and Ivan Pavlov, who proposed the simple reflex arc as the basic architecture of behavior.[2] His Pavlovian focus on automatic and involuntary behavior, in which stimulus always *preceded* response, had no place for reward or punishment coming *after* the action. The process by which the pleasurable consequences of an action somehow "stamped it in" seemed to him occult and implausible, and in a 1917 book he dismissed Thorndike's Law of Effect, suggesting that it was "a bit strange that scientifically-minded men should have employed it in an explanatory way."[3]

The hedonist baton was then accepted by the twentieth century's most infamous behaviorist, Burrhus Frederic Skinner. As a graduate student at Harvard, Skinner spent most of his working hours in the machine shop. An adept tinkerer and inventor since childhood, he developed a battery of technologies for measuring reward-seeking behavior, ending up with an "experimental arrangement not unlike the famous problem boxes of Edward L. Thorndike."[4] Skinner figured out how to make the rate at which rats helped themselves to food pellets self-recording, and showed that the resulting curves revealed mathematical regularities. Soon he began to develop his own vocabulary for the phenomena of reward-seeking behavior. First he introduced the word *reinforcement* for the pleasurable consequences of an act that increased the likelihood of its being repeated.[5] Later he coined the word *operant* for behavior that "acts upon the environment in such a way that a reinforcing stimulus is produced."[6] His technique for shaping behavior he named "operant conditioning." One of his most important discoveries was made by accident, when his food dispenser jammed and he found that the rat pressed the lever even more compulsively when it failed to produce a reward. Eventually he worked out how to stretch the reward ratio a little at a time, inducing pigeons to keep pecking thousands of times for a single grain of corn, like a gambler at a cannily programmed slot machine.[7]

After World War II Skinner began to conceive of his behavior control techniques as applicable to human life outside the laboratory, and he issued a series of hedonist manifestos urging social reform by means of positive reinforcement. The first of these was a utopian novel, published in 1948, describing an idyllic behaviorist community in the American heartland. Sales of *Walden Two* were disappointing, but Skinner's visibility grew with his 1948 move to Harvard, where he taught the principles of operant conditioning using live pigeons in the classroom. To make his point against the "mentalistic" vocabulary that he despised, Skinner trained pigeons to display what he called "synthetic social relations," getting them to play avian Ping-Pong against one another and to peck at discs in union in order to get food rewards. "Cooperation and competition were not traits of character," he concluded, but were "capable of being constructed by arranging special contingencies of reinforcement."[8] Despite the consistency with which pigeon behavior could be shaped in this way, classical behaviorists of the Watsonian-Pavlovian school criticized the whole notion of operant conditioning, asking how, exactly, a reward coming after an action strengthened the likelihood of that action's being repeated.

In 1953, however, the aura of mystery around the mechanism of reinforcement began to dissipate when two postdoctoral fellows in psychology at McGill University implanted electrodes in rat brains and put them in "Skinner Boxes." One of them had been running tests on the *aversive* effects of direct electrical stimulation of the limbic system, when a misplaced electrode seemed to have the opposite effect. Realizing that they had stumbled upon a possible pleasure center, they connected a lever in the cage to an electrode implanted in the septal area of the rat's brain and measured the rate at which the animal pressed the lever. They concluded that "the control exercised over the animal's behavior by means of this reward is extreme, possibly exceeding that exercised by any other reward previously used in animal experimentation."[9] With their serendipitous discovery of an area of the brain that rats seemed to want—desperately—to stimulate, they supplied operant conditioning with a plausible biological substrate. The experiment became one of the legends of psychological research, with the compulsively lever-pressing rat standing as a potent image of pleasure-seeking and addiction in humans.

## The Authoritarian Personality

SKINNERIAN BEHAVIORISM WAS by any standards a stunning success. Repeated invocations of the "laws of behavior" seemed to elevate psychology to the status of a hard science. The framework of operant conditioning opened up a vast field of possibilities for designing, executing, and publishing experiments, and training rats and pigeons became "highly reinforcing" for graduate students. Economists embraced behaviorism as congruent with their analyses of desire, utility, and choice, and its manipulative potential appealed to the more technocratic reformers of the Progressive Era.[10] But while the first decades of the century had witnessed the rise of a functionalist approach to social improvement, the aftermath of World War II gave the distinctly unbehaviorist ideal of human moral freedom a new lease on life.[11]

An epic volume was published in 1950, *The Authoritarian Personality*, a thousand-page psychoanalytic study of fascism and ethnocentrism sponsored by the American Jewish Committee. The goal was to understand the "social-psychological factors which have made it possible for an authoritarian type of man to threaten to replace the individualistic and democratic type prevalent in the past century and a half of our civilization and of the factors by which this threat may be contained."[12] Subjects were measured on the "F scale" for how latently fascist they were, on the "E-scale" for their ethnocentrism, and on the "AS Scale" for their anti-Semitism. High scorers on any of the scales turned out to be very unpleasant people indeed, prone to criminality, superstition, delinquency, and a variety of mental illnesses. The study was criticized for its Freudianism (high scorers were seeking "sadomasochistic resolution of the Oedipus complex"[13]) and for neglecting leftist totalitarianism in favor of the far-right variety. Despite these quibbles, work continued apace on the cluster of dispositions that constituted the antidemocratic type. One researcher, for example, administered the E-scale to a group of college students and then showed them a simple asymmetrical diagram. After removing the figure from view, he asked them to draw it from memory. Two weeks later he asked them to reproduce it again, and then again after four weeks. His graph of the results showed that "those high in ethnocentrism displayed an increasing tendency to reproduce the figure symmetrically." Just as the authors of *The Authoritarian Personality* had suggested, conformity not only had unpalatable political consequences; it also generated "lies and errors in vision, memory, or logic."[14]

Soon behaviorists began to come under fire for displaying just those qualities of rigidity and narrow-mindedness that characterized the authoritarian personality. A 1949 paper in *Psychological Review* accused behaviorism of having a "static, atomistic character," lacking creativity, neglecting questions of value, intensifying scientific hierarchies, and promoting the work of "technicians rather than discoverers."[15] At Skinner's home institution of Harvard, the anti-behaviorists founded the interdisciplinary Department of Social Relations, "identified by its interests in personality, social issues, and a commitment to the view that humans are naturally autonomous." Skinner and his mentors were left behind to rule over a Psychology Department defined by a "deterministic, behavioristic perspective on human nature."[16] As the two sides entrenched, anti-behaviorist critiques tended more toward the insultingly diagnostic, with an important 1959 survey of psychological science denouncing behaviorism's "attachment to a 'facile' mythology of perfection, as well as its 'autism.'"[17]

To its critics the behaviorist universe was devoid, above all, of *meaning*. There seemed to be no place in it for concepts, judgments, values, intentions, or beliefs. In 1948 the psychologist Karl Lashley gave a presentation titled "The Problem of Serial Order in Behavior" in which he proposed that behavior was dictated not just by the environment but also by the active organizational capacities of the "intact organism." The "problem of serial order" was best exemplified, of course, by syntactical language: successions of external stimuli could never account for the proverbial newsworthiness of "man bites dog." Taking up the gauntlet, Skinner published an audacious treatise titled *Verbal Behavior* (1957), in which he asserted that all linguistic phenomena could be analyzed in the terms of operant conditioning: the meaning of words was a function of the reward given for a certain behavior, such as looking at a chair, uttering "chair," and being rewarded by a parent's delighted hand clapping.

The book was received politely at first, until the linguist Noam Chomsky published a long, damning review, claiming that the theory of operant conditioning might contain valuable insights into the behavior of rats and pigeons but that extrapolating it to human language deprived it of all its precision. He advanced an alternative hypothesis, in which all human languages partook of a universal grammar, produced by the unique structure of the human brain and showing marked developmental stages.[18] The review soon began to attract more attention than the book itself, and in the years that followed, Chomsky helped develop a research program in psycholinguistics, devoted to probing the fundamental principles of operation of the human mind. Soon the new field was housed in its own institution: the Harvard Center for Cognitive Studies.[19]

## Studies in Schizophrenia

SKINNER'S RATS AND pigeons may have stood for everything that Cambridge liberals loathed, but far away from the Ivy League spotlight, at Tulane University in New Orleans, invasive hedonic research on human subjects was able to flourish unchallenged. The Tulane electrical brain stimulation program was directed by Robert Galbraith Heath, the son of a doctor, who had obtained his medical degree from the University of Pittsburgh in 1938. Because he had specialized in neurology, Heath was drafted into the army during World War II, given a two-month training in psychiatry, and thrown into the deep end, treating psychic war trauma.[20] After the war he continued his psychological and neurological training at Columbia University. There he came under the influence of an émigré Hungarian psychoanalyst, Sandor Rado, who was rewriting all the problems and theories of Freudian psychoanalysis—sex, defecation, death, dreams, anxiety, Oedipal desires, and so on—in terms of what he called "hedonic self-regulation" and evolutionary utility. Here is Rado on the joys of the table: "It is a very impressive arrangement of nature that an activity of such high utility value as ingestion should have such tremendous pleasure value for the organism."[21] Heath was spellbound, later describing Rado as "a brilliant person, extremely creative, very intuitive in his understanding," with "a more basic understanding of human behavior than any individual before or since, including his mentor Sigmund Freud."[22] It was from Rado that Heath internalized the evolutionary-hedonistic approach that would define his research agenda.

When Heath arrived at Columbia, a study with patients at the Greystone Hospital in New Jersey was under way, intended to bring some scientific rigor to the assessment of surgical treatments for mental illness. A group of forty-eight patients was selected, then sorted into twenty-four pairs with matching psychiatric characteristics. One member of each pair was then operated on, while the other served as a control. After surgery all the patients were subjected to the same rehabilitative efforts and psychological tests. Heath's job was the administration of the tests. The results of the study were not encouraging, but Heath was marginally more positive than most of his colleagues, suggesting that the removal of Brodmann areas 9 and 10 produced lasting improvements in some schizophrenic patients.[23]

Inspired by Rado's suggestion that the seat of the emotions was to be found in the deeper, more primitive parts of the brain, Heath began to argue that the frontal cortex was the wrong place to intervene in emotional disorders (in which category he included almost all forms of mental illness). Perhaps a better approach would be electrical stimulation of structures inaccessible to the surgeon's knife? He did not find any takers among his East Coast colleagues, but in 1949, before he had even completed his doctoral research, he was recruited by the dean of Tulane's medical school to head a new department of psychiatry and neurology. He later recalled that "no-one yet was ready to move into the subcortical structures . . . anywhere in the North," whereas New Orleans offered "real opportunities, because it was such a backward area. . . . There was no department, but we did have this vast institution called Charity Hospital. There was a tremendous amount of clinical material."[24]

Heath embarked on his research in 1950, aided by an interdisciplinary team of psychologists, psychiatrists, physiologists, surgeons, and clinicians. In the first round of treatments, twenty patients with "hopeless" schizophrenia were implanted with electrodes in a variety of deep brain structures. Also included were three patients with intractable pain from cancer, two with psychosis associated with tuberculosis, and one with severe rheumatoid arthritis, to serve as controls. The placement of the electrodes was based on Heath's conclusion that the removal of Brodmann areas 9 and 10 produced therapeutic improvements in the Greystone patients. Since that time he had conducted a series of animal experiments verifying that these areas were anatomically connected to the septal region and showing that lesions to this area of the brain in cats produced behavioral and metabolic changes analogous to schizophrenic symptoms in humans. For experimental purposes he also chose a few other sites for implantation. Believing that the wiring between the emotional centers and the cortex was scrambled in schizophrenics, he reasoned that electrical stimulation might restore the circuits to functioning.

In June 1952 Heath organized a three-day symposium to showcase the results. His final assessment of the treatment was wildly affirmative. According to a "Table of Therapeutic Results," only four of the twenty patients failed to show any improvement, two of whom were judged to be "technical failures" and so did not count. The rest were considered to have made progress, including five who had gone from "hopeless" before the treatment to "minimal defect" afterward. Some of the invited discussants were impressed; others less so. Responses ranged from the adulatory— "The change is so dramatic and so real that one is reminded of the fairy story of 'The Sleeping Beauty'"—to the skeptical: "No sound basis has been advanced here for the assumption that schizophrenia is due to specific septal pathophysiology or that this condition is influenced by manipulations of this region."[25]

But even the most skeptical respondents congratulated Heath for his courage in undertaking such an experiment. The introduction of the "chemical lobotomy" in the form of antipsychotic drugs was still in the very first stages in 1952, and so at the time of the symposium Heath's research represented relatively conservative therapeutic exploration.[26] It did not remain so for long. Even as the standard of care changed around him as a result of chlor-promazine and other drug treatments, Heath carried on experimenting with the electrical stimulation technique, during which time his theoretical justification underwent a subtle transformation. In the first phase he transposed a psychoanalytic view of the cause and progression of mental illness onto a dynamic model of the wiring of the brain. After 1953, inspired by the Canadian experiments on the pleasure centers of rats, his electrical stimulation of the septal area evolved into an attempt to shape human behavior with the ultimate neural reward. At the beginning of the 1960s he was once again ready to showcase his results, but this time he was more out of step with the wider world. Since antipsychotic drugs were now available, treating psychosis with invasive surgery of dubious therapeutic value looked much riskier than taking a pill. He also had the misfortune to preside over his second major symposium just as the image of the electrode-wielding behavioral scientist had become an emblem of cold war malevolence, courtesy of a social psychologist at Yale by the name of Stanley Milgram.

## Obedience to Authority

"TODAY," THE PREFACE to the 1950 study *The Authoritarian Personality* noted, "the world scarcely remembers the mechanized persecution and extermination of millions of human beings only a short span of years away."[27] As has often been pointed out, it was not until the beginning of the next decade that discussion of Nazi atrocities became widespread, spurred by the trial of Adolf Eichmann in Jerusalem. One of the people whose understanding of the Holocaust was crystallized by the Eichmann trial was Stanley Milgram, whose "obedience to authority" experiment transposed the liberal critique of behaviorism into a compelling image of a white-coated, affectless psychologist, armed with an electroshock machine.[28]

Milgram was the child of two Eastern European Jews who had come to the United States in the first decades of the twentieth century. In 1960 he completed a PhD dissertation probing the differences in "social conformity" between Norwegians and French people. The setup was simple: it recorded how many times a subject would give an obviously false answer to a question when he or she was tricked into believing that everyone else in the study had answered that way. These experiments were cast as a series of investigations into "national character" (Norwegians turned out to be more conformist than the French) but were also a manifestation of the free-ranging anxieties of the 1950s about organization men, authoritarian personalities, and the relationship of the individual to the crowd. Sometime in the spring of 1960, inspired by Eichmann's arrest, Milgram conceived of a sinister twist on his conformity research, asking if "groups could pressure a person into . . . behaving aggressively towards another person, say by administering increasingly severe shocks to him."[29] In October and November of that year he sent letters of inquiry to three government agencies about the prospects of grant support for his research into obedience. His students built a fake "shock box," and he ran some preliminary studies using Yale undergraduates as subjects.

The money duly rolled in, and the experiment went ahead. Subjects were told that they were participating in a study of punishment and learning, in which they and another participant would either play the role of "teacher" or "learner," to be determined by drawing lots. What they did not know was that the other participant was a confederate of the experimenter and that both of the pieces of paper said "teacher" on them. The two men were then treated to a brief theoretical lecture about the role of punishment in learning and memory, then the putative learner was led away and the real subject of the experiment was seated in front of the shock box. This was an authentic-looking piece of equipment with a series of thirty switches labeled from 15 to 450

volts, grouped into batches of four switches with captions running from SLIGHT SHOCK to SEVERE SHOCK, the last two simply and ominously labeled XXX. The teacher was then instructed
to administer a simple word-association test and to punish the
learner's wrong answers with shocks of increasing severity. The
subjects were unable to see the learner, but his cries of anguish
(prerecorded for consistency) were clearly audible. Famously,
about a third of participants continued with the experiment up
to the last set of switches.

Milgram claimed that his experiment was "highly reminiscent
of the issue that arose in connection with Hannah Arendt's book
*Eichmann in Jerusalem*. Arendt contended that the prosecution's
effort to depict Eichmann as a sadistic monster was fundamentally wrong, that he came closer to being an uninspired bureau-
crat who simply sat at his desk and did his job. After witnessing
hundreds of ordinary people submit to the authority in our own
experiments, I must conclude that Arendt's conception of the
*banality of evil* comes closer to the truth than one might dare imagine."[30] In other words, he argued that he had exposed the universal psychological trait of obedience linking ordinary Connecticut
folk to the defendant in Jerusalem.[31] For present purposes, what
is interesting about his experimental protocol is that the "authority" in the experiment was derived from the psychology of reward
and punishment. The subject was told that "psychologists have
developed several theories to explain how people learn various
types of material. . . . One theory is that people learn things cor-
rectly whenever they get punished for making a mistake."[32] Milgram updated the antitotalitarian ideal for the height of the cold
war: the banality of evil in his memorable performance piece was
not a fascist preoccupation with racial hierarchies and the purity
of the genetic pool but rather the dedicated environmental determinism of pleasure-pain research.

The decade that began with Milgram's infamous experiment
quickly generated other seminal critiques: the following year com-

munist brainwashing was depicted in the film *The Manchurian Candidate* and aversive operant conditioning portrayed in the novel *A Clockwork Orange*. In 1963 *Life* magazine ran an article on "chemical mind changers," which opened with a riff on the cranky, competitive founder of Skinner's fictional utopian community, "Frazier." Thereafter Skinner's name was repeatedly invoked in relation to debates about behavior modification. He responded that so-called mind control happened everywhere, all the time, through the perfectly ordinary processes of reward and punishment found in every natural and social environment. The only question that confronted the human species, he declared, was whether to design these environments scientifically or leave the whole business up to chance: "I merely want to improve the culture that controls."[33] As anxiety about the fascist personality was joined by paranoia about Stalinist brainwashing, the behavioral scientist became the emblem of an American version of totalitarianism.

### The Role of Pleasure in Behavior

IN 1962, ONLY a year after Milgram's study, Heath organized a second symposium at Tulane. Twenty-two scientists participated, including James Olds, the neuroscientist who had accidentally discovered the rat's pleasure center a decade earlier. Heath coauthored two chapters, the first of which, titled "Attempted Control of Operant Behavior in Man with Intracranial Self-Stimulation," recounted his efforts to replicate Olds's results in humans: "Thus far, reported self-stimulation work under controlled laboratory conditions has been confined to subhuman species. The present report described exploratory efforts in the extension of such studies to man." In July 1962 Heath's team had implanted depth electrodes in two human subjects, "a 35-year-old divorced white man with a diagnosis of chronic schizophrenic reaction, catatonic type," and "a 25-year-old single white man" suffering from "temporal lobe epilepsy with possible underlying schizophrenia."[34]

The experiment went badly. To the frustration of the research team, "when the current was turned off, both subjects continued to press the lever at essentially the same rate until the experiment was terminated by the experimenter after hundreds of unrewarded responses." When asked why they continued to press the lever, the "schizophrenic patient would invariably state that it felt 'good'; the epileptic subject would say that he was trying to cooperate with us, that he assumed we must want him to press it, since he had been placed there." The researchers' frustration with the recalcitrance of the second man began at this point to show: "Despite fairly forceful reminders in his case of the actual instructions, the same behavior recurred repeatedly, and the same sort of explanation was invariably given." The scientists had invested their brightest hopes in this patient because he was "intellectually intact," but he turned out to be impossible to work with: "Besides frequent sullen and demanding behavior, he feigned adverse reactions to the stimulation (even on occasion with the current turned off)." Most of the results were thus obtained from the other man, "B12," who presented problems of his own, as he "was catatonic, and might have sat for hours without making a movement that could be reinforced."[35]

Given the problems with getting the men to stop self-stimulating after the current was turned off, the experimenters introduced a new protocol, which they dubbed, without apparent irony, the "Free-Choice Procedure." By this time they were working with only B12, the catatonic patient, and they presented him with two different things to press, a lever and a button, hoping that if he were allowed to choose between them, he might select the rewarding stimulation. This produced mixed results. When the choice was between rewarding and aversive current, B12 chose the positive stimulation, but "most attempts to control such behavior under conditions of rewarding current versus no current were unsuccessful." Heath and his team tried injecting B12 with methamphetamine to enhance his sensitivity to the stimulation, but "the drug produced no detectable change in response rate."[36]

The paper's disarmingly frank conclusion was that "attempts to establish, modify, and extinguish a simple lever-pressing response under conditions of intracranial self-stimulation in two human subjects have proved largely unsuccessful." Heath and his coauthors nonetheless asserted that, "with revisions of procedure, data were obtained which suggest the presence of subcortical areas in the human brain in which brief electrical stimulation appears to have rewarding or reinforcing properties." There was no discussion of the possible confounding effect of using human subjects, able to talk back, to resist, to obey, and, as on this occasion, to obey robotically as an exquisitely annoying form of resistance. Instead Heath attributed the problems with the experiment to the patient's mental illness: "Since most of our findings are based on the intracranial self-stimulation behavior of one clearly non-normal subject, they should be interpreted with caution." [37]

## Beyond Freedom and Dignity

DESPITE THE SUSTAINED academic opposition to behaviorism, the 1960s was marked by growing enthusiasm for Skinner's work from a wider public outside the ivory tower. At the beginning of the decade sales of *Walden Two* began to rise; by decade's end a series of intentional communities had been founded, explicitly inspired by the book. Most of these ran into trouble when the communards discovered that Skinner's antidemocratic approach—in which planners and managers discretely ran the show and everyone else just enjoyed the results—was unworkable in the context of the American 1960s counterculture. It turned out that anyone predisposed to start or join a "Walden Two" either thought that he or she should be the Frazier figure or, more modestly, desired to share equally in the decision making. [38] Democratic modifications to strict behaviorist principles were mostly undertaken without relinquishing adherence to Skinner's larger vision, however, and by 1970 his talks were attracting huge crowds to college auditoriums.

In 1971 Skinner published a summary of behaviorist philosophy, aimed at a general audience, which stayed on the *New York Times* best-seller list for twenty weeks. The provocatively titled *Beyond Freedom and Dignity* laid out his blueprint for solving the "terrifying problems that face us in the world today," including population explosion, nuclear holocaust, famine, disease, and pollution. The humanist, he explained, appealed to mentalistic explanations such as "attitudes," "opinions," and "personalities" to account for self-destructiveness; his aim was to replace the language of internal mental states with unadorned descriptions of human behaviors and the environmental conditions that determined them. That year he made forty radio and television appearances.[39]

Skinner's principal target was a cognitive specter he called "Autonomous Man," an Enlightenment superstition much beloved by the authors of "the literatures of freedom." "Autonomous man," he scoffed, "serves to explain only the things we are not yet able to explain in other ways." "Responsibility" was another superstition that Skinner attributed to our ignorance of the true determinants of behavior: "The real issue is the effectiveness of techniques of control. We shall not solve the problems of alcoholism and juvenile delinquency by increasing a sense of responsibility. It is the environment which is 'responsible' for the objectionable behavior." These metaphysical specters were impeding "the design of better environments rather than of better men."[40]

Despite his avowed commitment to solving the great problems of the day, Skinner made the most strenuous efforts to present his work as value-free. "Good things are positive reinforcers. The food that tastes good reinforces us when we taste it," he announced. Having collapsed the good and the pleasurable, he was then able to translate moral axioms into a behaviorist idiom: "If you are reinforced by approval of your fellow men, you will be reinforced when you tell the truth." In the same key he averred that his behaviorist prescriptions concerning pollution, population, and nuclear war were not predicated on any vision of the good life but rather on his evolutionarily programmed and culturally reinforced desire for the survival of his species and of his culture.[41] His determination to strip even human happiness of moral content left him bereft of substantive reasons to enact his program of social reform. His supporters and followers outside the academy wanted desperately to build a better world, but his scientism had thrown him back on a species of nihilism that would prove unequal to the moral passions of those avatars of freedom, dignity, and autonomy converging in ever-greater numbers on the antipsychiatry movement.

## Zero plus Zero

CHOMSKY'S REVIEW OF *Beyond Freedom and Dignity* in the *New York Review of Books* made his earlier review of *Verbal Behavior* look like the model of academic politeness. Chomsky asked what the book contributed to psychology and concluded that "zero plus zero still equals zero." As for Skinner's proposals for societal reform, he summoned up the vision of "a well-run concentration camp with inmates spying on one another and the gas ovens smoking in the distance." He admitted that "it would be improper to conclude that Skinner is advocating concentration camps and totalitarian rule," but only because this "conclusion overlooks a fundamental property of Skinner's science, namely, its vacuity."[42]

It was despair and anger over the Vietnam War that had sent Chomsky and his fellow travelers to such rhetorical extremes, dividing the world into those who opposed the fascism of U.S. foreign policy and those who collaborated with it. In his 1959 review of *Verbal Behavior*, Chomsky's exemplar of the non-behaviorist human was an unworldly scholar; in the 1971 review it was an antiwar activist: "Suppose that a description of a napalm raid on a foreign village induces someone in an American audience to carry out an act of sabotage. In this case, the 'effective stimulus' is not a reinforcer, but the mode of changing behavior may be quite effective." Another reductio ad absurdum of Skinner's argument linked him to Eichmann and the architects of the Vietnam War. A chapter in *Beyond Freedom and Dignity* argued that we are more likely to give people credit for actions whose behaviorist motivations we do not understand, such as "those who live celibate lives, give away their fortunes or remain loyal to a cause when persecuted, because there are clear reasons for behaving differently."[43] Chomsky parsed this with deliberate obtuseness: "Skinner is claiming that if Eichmann is incomprehensible to us, but we understand why the Vietnamese fight on, then we are likely to admire Eichmann but not the Vietnamese resistance."[44] The review was considered a major event in the world of letters, and 1971 was the year that witnessed the steepest decline in articles and PhD dissertations on behaviorist psychology and a corresponding sharp rise in the output of Chomsky's cognitive school.[45]

Chomsky's hatchet job showed how prophetic Milgram had been. The 1961 Obedience to Authority experiment had taken the analysis of *The Authoritarian Personality* of the 1950s, radicalized it, and recast it in a psychiatric idiom perfectly suited to the Vietnam War era. Recognizing no moral distinction between My Lai and the Nazi invasion of Poland, the antiwar movement turned the liberal critique of totalitarianism against liberalism itself, with the behaviorist psychologist standing as the emblem of the unfeeling machinery of the American state and the academic-industrial complex that supported it.

Activists began to attend Skinner's public lectures, including a group calling themselves the National Caucus of Labor Committees, who announced that they had "delivered a Nuremberg Indictment for Crimes Against Humanity" against "the nation's number one populizer [*sic*] of Nazi medicine, B. F. Skinner." On the campus of Indiana University they hanged him in effigy. At one talk they fired a blank cartridge during the question-and-answer session. Police began to stand by at Skinner's public appearances, and he made sure to walk to his office at Harvard by a different route each day.[46] In 1974 Skinner published a defense of his philosophy titled *On Behaviorism*. Reviewing the book, the radical psychologist Thomas Szasz wrote, "I believe that those who rob people of the meaning and significance they have given their lives kill them and should be considered murderers, at least metaphorically. B. F. Skinner is such a murderer. Like all mass murderers, he fascinates—especially his intended victims."[47]

## Septal Stimulation for the Initiation of Heterosexual Behavior

MEANWHILE HEATH WAS still working out the therapeutic possibilities for direct electrical stimulation of the brain. In 1970 he embarked on an attempt to "employ pleasure-yielding septal stimulation as a treatment modality for facilitating the initiation, development, and demonstration of . . . heterosexual behavior in a fixed, overt, homosexual male." This experiment betrayed the lasting influence of Rado, whose emphasis on the evolutionary utility of pleasure had resulted in a theory of sexuality that was far more prescriptive than Freud's: "I know of nothing that indicates that there is any such thing as innate orgiastic desire for a partner of the same sex. . . . The homosexual male often clings to the myth that he belongs to a third sex, superior to the rest of mankind. This would seem to be the effort of an individual who lives in constant dread of detection and punishment, which is the milieu of the society that prohibits homogenous mating, to restore his shaky equilibrium."[48] Partly under the influence of Rado's theory of maladaptive sexuality, reparative therapy enjoyed a vogue in the midcentury, with psychiatrists prescribing talk therapy, hormones, drugs, and other treatments to restore proper sexual functioning. In this context electrical brain stimulation seemed to promise a new horizon of efficacy.

Although Heath and his coauthor chose to foreground this eye-catching aspect of their treatment, the subject's sexuality was only one among many maladaptive aspects of his behavior that they set out to reprogram. According to their account, patient "B19" had an abusive father, a cold mother, and a long record of social isolation and behavioral problems, culminating in his being discharged from the military and spending two years as a "'drifter' travelling idly around the country, engaging in numerous homosexual relationships and being supported financially by homosexual partners." The list of symptoms included "distinct preoccupation with his body image . . . extreme somatization . . . hypochondriacal traits . . . paranoid ideation . . . apathy, chronic boredom, lack of motivation to achieve and a deep sense of being ineffectual, inadequate, worthless and inferior." EEG readings seemed to indicate that he suffered from temporal lobe epilepsy.[49]

B19 was put under general anesthetic and electrodes were implanted in eight deep brain structures. Four weeks later he was made to watch a heterosexual "'stag' film," at the end of which "he was highly resentful, angry, and unwilling to respond." The following day he embarked on a schedule of stimulation of the septal region, sometimes administered by the researchers, sometimes self-administered. He "exhibited improved mood, smiled frequently, stated that he could think more clearly, and reported a sense of generalized muscle relaxation. He likened these responses to the pleasurable states he had sought and experienced through the use of amphetamines." Eventually "he had to be disconnected, despite his vigorous protests." Most important, he "reported increasing interest in female personnel and feelings of sexual arousal with a compulsion to masturbate," after which "he agreed without reluctance to re-view the stag film, and during its showing became sexually aroused, had an erection, and masturbated to orgasm."[50]

Over the next few days B19 reported "continued growing interest in women." After another series of septal stimulations, he expressed "a desire to attempt heterosexual activity in the near future." The researchers accordingly arranged for a twenty-one-year-old prostitute to have sex with him in a laboratory while he was hooked up to an EEG apparatus to measure his brain activity. She seems to have been most therapeutically adept, allowing him to talk to her for an hour about "his experiences with drugs, his homosexuality and his personal shortcomings and negative qualities," during which she was "accepting and reassuring." In the second hour, "in a patient and supportive manner, she encouraged him to spend some time in a manual exploration and examination of her body." They had sex, and, "despite the milieu and the encumbrance of the electrode wires, he successfully ejaculated." Heath declared the experiment a success: "Of central interest in the case of B19 was the effectiveness of pleasurable stimulation in the development of new and more adaptive sexual behavior."[51]

For a protégé of Rado, it made sense to lump in homosexuality with depression, paranoia, grandiosity, and suicidal tendencies as just another evolutionarily maladaptive disorder of hedonic regulation. When the study was under way, however, a campaign by gay activists succeeded in getting homosexuality removed from the *Diagnostic and Statistical Manual of Mental Disorders*, the so-called Bible of psychiatry. The seventh printing of the *DSM-II*, published in 1974, contained a new diagnosis, "Sexual Orientation Disturbance (SOD)," suffered by "individuals whose sexual interests are directed primarily toward people of the same sex and who are either disturbed by, in conflict with, or wish to change, their sexual orientation." By pivoting around lack of self-acceptance, this diagnostic category implied that treatment for SOD might consist of a few affirmative sessions with a gay therapist rather than any sort of procedure aimed at changing the sexual orientation of the patient.[52] SOD replaced the highly paternalistic stance of reparative therapy (in Heath's case anchored in evolutionary theory) with an ethic of choice and self-determination.

## Human Experimentation and the Triumph of Autonomy

BY THIS TIME medical and psychiatric paternalism was under attack from all sides. The rewriting of the diagnosis of homosexuality coincided with the revelation of the Tuskegee Study of Untreated Syphilis in the Negro Male, a U.S. Department of Public Health initiative dating from 1932, in which 399 African American share-croppers diagnosed with syphilis had been selected for a study of the untreated disease. In 1972 a DPH employee told a journal-ist friend about the research, and on July 26, 1972, the study was on the front page of the *New York Times* under the headline "Syphilis Victims in U.S. Study Went Untreated for 40 Years."[53] The next day the paper ran an interview with a survivor—the grandson of slaves—who recalled that the researchers recruited subjects on the pretext that they were getting free treatment. In the same article a researcher from the Centers for Disease Control described the study as "almost like genocide."[54] On July 30 the paper's science writer noted that the study was "begun the year Hitler came to power. It was Hitler's atrocious 'experiments' done in the name of medical science which led after World War II to the promulgation of the Nuremberg Code."[55]

In response Senator Edward Kennedy organized a series of congressional hearings on the problem of human experimenta-tion, which were held over the course of six days between the end of February and the beginning of March 1973. Day three of the hearings was devoted to the neurosciences. The radical psychotherapist Peter Breggin delivered a sermon against the "mechanistic, anti-individual, anti-spiritual view," which "gives justification to the mutilation of the brain and the mind, in the interests of controlling the individual." For Breggin the root of the problem lay in the "totalitarian" outlook of psychologists such as Skinner, whose work, he asserted, "ridicules the basic American values: Love of the individual, love of liberty, personal responsi-bility, and the spiritual nature of men. . . . If America ever falls to totalitarianism, the dictator will be a behavioral scientist and the secret police will be armed with lobotomy and psychosurgery."[56]

After Breggin, Heath took the stand and presented the results of his experiments, including showing some film of his subjects undergoing stimulation. The first film showed a patient "in whom we turned on the adversive brain circuitry to induce violent impulses." The second film depicted "pleasure sites of the brain [being] stimulated to relieve physical pain" and in the third "pleasure sites [being] stimulated to remove the emotional pain of episodic rage and paranoia." After watching the films, Kennedy attempted to clarify the nature of the intervention: "What you are really talking about is controlling behavior." Heath bridled at this: "I am a physician and I practice the healing art. I am interested in treating sick behavior—not in controlling behavior." Undaunted Kennedy went on: "You have shown and testified about how you can replicate pain and pleasure by the implantation of these electrodes in different parts of the brain." This time Heath assented. Kennedy continued, "This is behavioral control. As I understand it, you are trying to use this technique to treat people." Again Heath agreed. Kennedy then asked, "Would it not be adaptable to treat other people as well, normal people?," to which Heath replied, "I think it would be, but I think normal adaptive people are already being treated. I am sure Dr. Skinner is going to talk on that. Our learning experiences, our attitudes are modified every day."[57]

Shortly afterward Skinner took the floor and elaborated on the continuity between Heath's electrical stimulation of the brain and his own, less invasive methods. "The control of human behavior through drugs, psychosurgery, or electrical stimulation naturally attracts attention," he observed, "but far more powerful methods have been in existence as long as the human species itself and have been used throughout recorded history." According to Skinner, there was no need for technological or chemical intervention, as "behavior is selected and strengthened by its consequences—by what the layman calls rewards and punishments, and this fact has long been exploited for purposes of control." Control, Skinner explained, was a fact of life: "We are all controlled all the time—by what has happened to us in the past and our social environment. The prisoner of war who resists efforts to demean him or change his views is not demonstrating his own autonomy; he is showing the effects of earlier environments—possibly his religious or ethical education, or training in techniques of resistance in the armed services."[58] In repudiating the notion of autonomy Skinner made clear the contrast between his view of humanity and the idea of the free, sovereign, individual human spirit that had motivated his opponents. But he was now up against a broader and more united opposition than at any time in his career: the cognitive model of human autonomy was identical with the ideal of moral freedom that underlay the movement for informed consent in medicine. By the end of the 1970s informed consent had become an absolute regulatory mandate, and the principle of patient autonomy had triumphed over the last vestiges of medical paternalism. Behaviorism remained an important presence in laboratory studies of learning in nonhuman animals, but it had to relinquish all of its larger philosophical and social ambitions.

As for Heath, after the hearings he hung on at Tulane, protected by his cloistered position within a private university, but he became the target of strident student protests. The Medical Committee for Human Rights, founded in 1964 to provide treatments for wounded civil rights activists, protested his public appearance in downtown New Orleans. In 1974 a local journalist published an account of his work, arguing that his flimsy consent procedures gave him "carte blanche permission for implantation, surgery, drugs and other treatments."[59] The article was titled "The Mysterious Experiments of Dr Heath: In Which We Wonder Who Is Crazy and Who Is Sane," an inversion of the categories of madness and reason that was an expression of the baffled fury of the activist generation that had grown up in the shadow of the Holocaust and discovered that the American victory over fascism was neither clean nor clear. Just as with Skinner, Heath's commitment to extreme behavioral plasticity had become associated with totalitarian mind control. Cognitive science, by contrast, with its human exceptionalism, its focus on reasoning and computation, its emphasis on information and symbolic language, was able to hitch its star to the democratic agenda.

In the complex welter of cold war politics, human moral freedom was possible only if neural plasticity was limited. In repudiating behaviorism the architects of cognitive science invoked the innate, genetically programmed capacities of the organism. In his review of Skinner's *Verbal Behavior* Chomsky suggested that "prediction of the behavior of a complex organism (or machine) would require, in addition to information about external stimulation, knowledge of the internal structure of the organism, the ways in which it processes input information and organizes its own behavior. These characteristics of the organism are in general a complicated product of inborn structure, the genetically determined course of maturation, and past experience."[60] For Chomsky and his fellow travelers it was this whole complex of traits that structured the responses of an organism to its environment. As linguists they believed the "genetically determined course of maturation" was especially salient. Language acquisition—the quintessential human capacity—seemed to unfold in an orderly, sequential fashion, displaying marked developmental stages. And just as Lashley had pointed out in 1948, semantic meaning could never be accounted for in a stimulus response framework; it must be an integrated capacity of the "intact organism" (i.e., not a decapitated frog), prior to any stimulus, that shaped patterns of behavior, especially verbal behavior.

By focusing on information processing as a computational problem, cognitive science neatly sidestepped the whole problem of organicism and genetic determinism that might have given their work eugenic overtones. What ultimately triumphed was the implied relationship between computational rationality and autonomous decision making, an Enlightenment framework updated for the computer age. This link between freedom and reason was perhaps best embodied in the empowered patient of informed consent, but it was equally compatible with the *Homo economicus* of neoliberalism, two iconic idealizations of human agency in the post–cold war world.

Meanwhile the physiologist Eric Kandel was busy studying the classical Pavlovian phenomena of habituation and sensitization in sea slugs. Administering electrical shocks to his animals as a proxy for learning and memory, he ultimately unlocked the sub-molecular mechanisms of neural plasticity, work for which he won the 2001 Nobel Prize. So, now that neuroplasticity is back, what *will* we do with this malleability? Can a neural process elucidated by means of Pavlovian conditioning truly represent liberation from the imperatives of biological determinism? Just as with epigenetics, it certainly seems to delineate some kind of freedom. But whether neuroplasticity represents freedom, flexibility, or Catherine Malabou's "explosiveness," do we want to hang on to any part of the Chomskian ideal of *structure*—meaning both the structured material substrate of the human brain's computational and symbolic capacities as well as the structure of reasoned argument—as ground for the possibility of democratic participation or political resistance?

# Endnotes

1   John B. Watson, "Psychology as the Behaviorist Views It," *Psychological Review* 20, no. 2 (1913): 158.

2   I. M. Sechenov, *Reflexes of the Brain* (Cambridge, MA: MIT Press, 1965), first published in 1863, does contain a brief discussion of pain and pleasure, but he attributes little behavioral significance to the primordial dichotomy (see 23–24).

3   John B. Watson, *Behavior: An Introduction to Comparative Psychology* (New York: H. Holt, 1914), 257.

4   B. F. Skinner, *The Shaping of a Behaviorist* (New York: Knopf, 1979), 87, 233.

5   Ibid., 97, 143.

6   B. F. Skinner, *The Behavior of Organisms* (New York: Appleton, Century, Crofts, 1938), 22.

7   C. B. Ferster and B. F. Skinner, *Schedules of Reinforcement* (East Norwalk, CT: Appleton-Century-Crofts, 1957), chapter 7.

8   B. F. Skinner, *A Matter of Consequences* (New York: Knopf, 1983), 25–26.

9   J. Olds and P. Milner, "Positive Reinforcement Produced by Electrical Stimulation of the Septal Area and Other Regions of Rat Brain," *Journal of Comparative and Physiological Psychology* 47, no. 6 (1954): 426.

10  John A. Mills, *Control: A History of Behavioral Psychology* (New York: New York University Press, 1998), 23–54, quotation on 30.

11  Jamie Cohen-Cole, *The Open Mind: Cold War Politics and the Sciences of Human Nature* (Chicago: University of Chicago Press, 2014).

12  Theodor W. Adorno, Else Frenkel-Brunswik, Daniel J. Levinson, and R. Nevitt Sanford, *The Authoritarian Personality* (New York: Harper, 1950), x.

13  Ibid., 759.

14  Cohen-Cole, *The Open Mind*, 43.

15  Daniel Brower, "The Problem of Quantification in Psychological Science," *Psychological Review* 56, no. 6 (1949): 325–33.

16  Cohen-Cole, *The Open Mind*, 85.

17  Ibid., 149.

18  Noam Chomsky, "A Review of Verbal Behavior by B. F. Skinner," *Language* 35, no. 1 (1959): 26–58.

19  Cohen-Cole, *The Open Mind*, 169–74.

20  Wallace Tomlinson, "Interview with Robert Galbraith Heath," video, Internet Archive, March 5, 1986, https://archive.org/details/WallaceTomlinsonInterviewingRobertHeath_March51986 (accessed June 4, 2015).

21  Sandor Rado, Jean Jameson, and Henriette Klein, *Adaptational Psychodynamics* (Northvale, NJ: J. Aronson, 1995), 57.

22  Tomlinson, "Interview with Robert Galbraith Heath."

23  Fred Mettler, ed., *Selective Partial Ablation of the Frontal Cortex* (New York: Paul B. Hoeber, 1949), 392.

24  Tomlinson, "Interview with Robert Galbraith Heath." Heath was wrong about the impossibility of doing electrical stimulation work on the East Coast, as witnessed by the success of Yale neuroscientist José Delgado, whose experiments on patients from a Rhode Island psychiatric hospital began at the same time as the Tulane research.

25  Robert G. Heath, ed., *Studies in Schizophrenia: A Multidisciplinary Approach to Mind-Brain Relationships* (Cambridge, MA: Harvard University Press, 1954), 502, 535.

26  Heinz Lehman and Thomas Ban, "History of the Psychopharmacology of Schizophrenia," *Canadian Journal of Psychiatry* 42 (1997): 152–62.

27  Adorno et al., *The Authoritarian Personality*, v.

28  Stanley Milgram, *Obedience to Authority: An Experimental View* (New York: Harper and Row, 1974).

29  Thomas Blass, *The Man Who Shocked the World: The Life and Legacy of Stanley Milgram* (New York: Basic Books, 2004), 62.

30  Ibid., 269.

31  Ian Nicholson, "'Shocking' Masculinity: Stanley Milgram, 'Obedience to Authority' and the 'Crisis of Manhood' in Cold War America," *Isis* 102, no. 2 (2011). This is the most historically contextualized treatment of Milgram's experiment, but it does not explore the critical implications of the behaviorist setup.

32  Blass, *The Man Who Shocked the World*, 77.

33  Skinner, *A Matter of Consequences*, 242.

34  Robert G. Heath, *The Role of Pleasure in Behavior: A Symposium by 22 Authors* (New York: Hoeber-Harper, 1964), 55, 57.

35  Ibid., 64–66, 62, 63.

36  Ibid., 73.

37  Ibid., 78, 79.

38  The notable exception was a behaviorist community in Mexico, under the strong leadership of a charismatic family patriarch. Hilke Kuhlmann, *Living Walden Two: B. F. Skinner's Behaviorist Utopia and Experimental Communities* (Urbana: University of Illinois Press, 2005).

39  Skinner, *A Matter of Consequences*, 317–21.

40 B. F. Skinner, *Beyond Freedom and Dignity* (London: Penguin, 1988), 36, 20, 77, 83. The supreme irony of including Mill in the roster of freedom's naïve and superstitious defenders was that he was arguably the first person to actually have been raised according to behaviorist principles.

41 Ibid., 103, 112, 124–42.

42 Noam Chomsky, "The Case against B. F. Skinner," *New York Review of Books*, December 30, 1971, http://www.nybooks.com/articles/archives/1971/dec/30/the-case-against-bf-skinner/ (accessed March 25, 2013).

43 Skinner, *Beyond Freedom and Dignity*, 50–51.

44 Chomsky, "The Case against B. F. Skinner."

45 R. W. Robins, S. D. Gosling, and K. H. Craik, "An Empirical Analysis of Trends in Psychology," *American Psychology* 54, no. 2 (1999): 117–28.

46 Skinner, *A Matter of Consequences*, 351–52.

47 Thomas Szasz, "Against Behaviorism: A Review of B. F. Skinner's *About Behaviorism*," *Libertarian Review* 111, no. 12 (1974): 6.

48 Rado et al., *Adaptational Psychodynamics*, 210–11.

49 Robert G. Heath and Charles E. Moan, "Septal Stimulation for the Initiation of Heterosexual Behavior in a Homosexual Male," *Journal of Behavioral Therapy and Experimental Psychiatry* 3, no. 1 (1972): 24–25.

50 Ibid., 26–27.

51 Ibid., 27–28, 29.

52 Robert L. Spitzer, "The Diagnostic Status of Homosexuality in DSM-III: A Reformulation of the Issues," *American Journal of Psychiatry* 138, no. 2 (1981): 211.

53 Jean Heller, "Syphilis Victims in U.S. Study Went Untreated for 40 Years," *New York Times*, July 26, 1972.

54 James T. Wooten, "Survivor of '32 Syphilis Study Recalls a Diagnosis," *New York Times*, July 27, 1972.

55 Jane E. Brody, "All in the Name of Science," *New York Times*, July 30, 1972.

56 U.S. Senate, Committee on Labor and Public Welfare, Subcommittee on Health, *Quality of Health Care—Human Experimentation, 1973: Hearings, Ninety-third Congress, First Session, on S. 974* (Washington, DC: U.S. Government Printing Office, 1973), 358.

57 Ibid., 367–68.

58 Ibid., 371.

59 Bill Rushton, "The Mysterious Experiments of Dr. Heath, in Which We Wonder Who Is Crazy and Who Is Sane," *Courier* (New Orleans), August 29–September 4, 1974.

60 Chomsky, "A Review of Verbal Behavior by B. F. Skinner," 27.

*Nima Bassiri*

# 4 Epileptic Insanity and Personal Identity: John Hughlings Jackson and the Formations of the Neuropathic Self

## Personhood and Pathology

A SPAN OF two hundred years, from the end of the seventeenth to the end of the nineteenth century, separates a striking reversal of positions on the relationship between madness and personhood. In 1694 John Locke published the second edition of *Essay concerning Human Understanding*, with the newly included chapter "Of Identity and Diversity"—Locke's expanded discussion of "personal identity."[1] In that chapter Locke grounded self-identity, the identity of being a single person, on the continuity of consciousness alone, the limits of which were defined by how far a person's consciousness of her thoughts, perceptions, and memories could extend. This approach lent to Locke's notion of personhood a structure of accountability, so that given the continuity of a singular experience, a person would ultimately be answerable at the very least to herself, morally and epistemically. Only then could she be answerable or publicly accountable to an institution or the law. Person or self, as Locke famously concludes, "is a Forensick Term, appropriating actions and their merit."[2]

For Locke the unity and singularity of personhood was its central characteristic, and to think otherwise was to stray toward an untenable position: "*If it be possible* for the same man to have distinct incommunicable consciousness at different times, it is past doubt the same man would at different times make different Persons; which, we see, is the Sense of Mankind in the solemnest Declaration of their Opinions, Humane Laws not punishing the Mad Man for the Sober Man's Actions, nor the Sober Man for what the Mad Man did, thereby making them two Persons."[3] Locke is, of course, not endorsing the idea that in madness we see the *authentic* emergence of a second person. In the case of madness we speak only *as if* the madman has become another self. This kind of figuration, that two selves inhabit a single man, does not amount to a viable claim about personal identity, which is why Locke begins this passage as a hypothetical supposition. Mental departures associated with madness properly represent nothing more than the temporary lacunae of the sober man's unified and conscious self. After all, Locke's broader concern is in demonstrating the absurdity that arises when personal identity is not grounded on consciousness but rather on *substance*—in particular, a Cartesian soul, a substance that perpetually thinks even when we are not cognizant of such thinking. Cartesianism, or some variation thereof, ensures that a person cannot be consciously accountable for all of her thoughts, insofar as she (or, rather, her soul) thinks even when she (that is, her conscious self) does not think she thinks. This not only upsets any possibility of self-accountability, but it also inevitably divides a person into two: a regularly thinking soul as well as an occasionally conscious self.

Exactly two hundred years later, however, another English physician will posit a near-reversal of this position, by not only asserting the *possibility* that in madness another self can authentically emerge (for Locke the position was not impossible, merely invalid and absurd) but also affirming its *acceptability* as an epistemologically and medically viable claim. In 1894 the seminal neurologist John Hughlings Jackson published an article titled "The Factors of Insanities" in the *Medical Press and Circular*. The essay represented a culmination of his most mature theories of the physiology and pathology of the brain and nervous system, further expounding the conceptual innovation for which he was most famous, namely his "evolutionary" understanding of the nervous system—a conception that the entire nervous system functionally represented the entire body, in an ever increasing hierarchy of complexity and specialization. The "lowest" centers of the peripheral nervous system represented the sensorimotor dimensions of the body, which were then re-represented (and re-re-represented) in higher cerebral centers according to a cumulative degree of differentiation and specification. The highest cerebral centers therefore did not localize a mind per se but rather the most specific, differentiated, novel, and complex sensorimotor coordinations.

Within this hierarchical and vertically layered picture of the evolutionary nervous system, neuropathology for Hughlings Jackson amounted to instances of "dissolution," or the functional cessation of a higher center combined with the consequent disinhibition of a lower function. Behavioral anomalies characteristic of madness were not simply expressions of the loss of higher processes but also the exhibition of the newly disinhibited lower centers. Such behavioral anomalies might include any number of "insanities" and altered states of consciousness, including automatism, somnambulism, dreamy states, and hallucinatory episodes.

In "The Factors of Insanities," however, Hughlings Jackson claimed that what was in fact exhibited in neural "dissolution" was not only madness but a state of consciousness so altered that one might rightly define it as a "new person," a person who is epistemologically and forensically distinct from the former, "sane" self. Madness authentically meant to become another person entirely. After all, the functions that were disinhibited during madness, while "lower," were nevertheless *normal* from both a physiological and an evolutionary standpoint. Pathology amounted to a shift or drop from one normal brain process to another—in effect the introduction of another brain and, so, another self.[4]

Between Locke and Hughlings Jackson, the discussion pertaining to the feasibility of understanding madness as a genre (rather than lacuna) of personhood had transitioned from philosophical absurdity to medical legitimacy. For Hughlings Jackson it had become medically viable and acceptable to speak in a certain way about personhood and madness—but only insofar as madness could be characterized in neurological terms.

What is noteworthy, however, is not the novelty of Hughlings Jackson's claims. Decades earlier, between 1822 and 1826, Franz Gall published his multivolume *On the Functions of the Brain*, in which he declared, "There are cases, where, by an alteration of the [cerebral] organs, the ME is transformed into another ME; for instance, when a man believes himself transformed into a woman, a wolf, etc.; there are cases, where the old ME is entirely forgotten or replaced by a new one; not an uncommon accident after severe disease, especially in cerebral affections."[5] But Gall's claims were not, as they would be for Hughlings Jackson, situated within a medically acceptable style of reasoning. This was, in other words, not yet a generally viable way of speaking about personhood and brain. Gall's assertions functioned more than anything as a conjectural rebuttal against a dominant eighteenth-century tendency to correlate the unity of the mind with what was viewed as the homogeneity of the brain's white matter.[6] Instead of a unified mind subtended by a homogeneous brain—a medical position that Gall argued adhered to philosophical commitments—Gall introduced a composite brain that could support a mosaic self (a challenge therefore to philosophy's reputed primacy over medical knowledge).

On the other hand, while Hughlings Jackson's claims regarding the neuropathology of personhood were somewhat unique, what makes them so noteworthy is that they were embedded within a set of psychiatric and neurological discussions, disputes, and controversies concerned at various levels with the status of "personal identity, and its morbid modifications"—to quote the title of a 1862 article by Hughlings Jackson's colleague, the psychiatrist James Crichton-Browne (along with Hughlings Jackson, one of the founding editors of the journal *Brain*).[7] Not only had it become important to inquire into the status of the self in states of madness and neurological disorder; it had also become necessary to consider that some state of personhood *could* be retained, or even newly fashioned, in pathological circumstances.

The discourse of behavioral medicine between 1860 and 1900, within which Hughlings Jackson's work must be broadly contextualized, comprises what Ian Hacking calls a "space of possibilities" enabling certain formations to emerge, formations in this case specifically related to the relationship between personhood, pathology, and the brain.[8] Not only is Hughlings Jackson representative of some of the most significant conceptual renovations to neurophysiology at the end of the nineteenth century; more important for the purposes of this essay, he is representative of a way in which personhood was coming to be neurologically "made up" by the end of the nineteenth century, and in such a way where person and brain became so conceptually linked that modifications to neural processes corresponded not to the loss but rather to the *modification* of self.[9]

Yet, as I will propose, there is more to the story of Hughlings Jackson's neuropathology of self than the suggestion that neurological paradigms historically opened up new possibilities for reimagining personal identity, which they certainly did. That standpoint, taken on its own, implies that brain research *alone* enabled such possibilities to arise. What I instead propose is that neurological discourse—Hughlings Jackson's in particular—only formalized and stabilized a *problem* related to the category of selfhood and the status of personal identity that was emerging as a consequence of broader discussions throughout behavioral medicine. Hughlings Jackson's proposition of pathologically becoming a new person functioned as a reflection of and rejoinder to a set of epistemological and forensic anxieties related to personhood itself.

## Normal and Pathological

MUCH OF THIS story revolves around a gradual transformation in the medical dichotomy of health and disease, the shifting relationship of which was profoundly exemplified in historical accounts of neuropathology. The pathological dissolution of the self, for instance, was not merely a seventeenth-century *philosophical* premise; it had its correlates in important early modern principles of neuroanatomy. For the seminal seventeenth-century neuroanatomists, the brain's anatomical organization was likened to and imagined through the architecture of political absolutism— quite literally a castle or fortress that housed and safeguarded a sovereign mind or soul. Thomas Willis and Humphrey Ridley were quite explicit in their assertion that the brain was "like a castle," where the cortical layers (viewed as little more than a functionless crust) were designed for the purposes of defense and safety against foreign threats.[10]

Indeed disease was understood as arising from an *external* morbidity or contagion, a breaching of the castle walls, and a toppling of the sovereignty of the mind itself. As Descartes confirms in a letter to Princess Elizabeth, "The architecture of our bodies is so thoroughly sound that when we are well we cannot easily fall ill except through extraordinary excess or infectious air or some other *external cause*."[11] If we were to believe Descartes when he proclaims, "The preservation of health has always been the principal end of my studies,"[12] then it would make sense for him to have located the pineal gland, the seat of the soul, at the exact center of the brain. Because there "it is situated in such a well-protected place that it is almost immune from illness."[13]

The external threat of disease was thoroughly incompatible with the stability of a healthy brain. Illness was framed as a rupture and penetration of the brain's coherence, potentially deposing whatever mind, soul, or self would be said to reside there. And as Hobbes warned, the body politic can no more be governed by an expired sovereign than can "the Carcasse of a man, by his departed (though Immortal) Soule."[14] Pathology was the undoing of health and the unseating of the rational mind to which health was inextricably linked. A diseased brain was a dethroned mind, the absence of the self altogether.

A more reconciled view of the relationship between health and disease, in the form of mutual compatibility, begins to emerge by the middle of the eighteenth century, at a moment when the life sciences witnessed a general turn toward vitalist approaches to the study of living beings. In his textbook *Institutions of Medical Pathology* (1758) Leiden medical professor Jerome Gaub viewed disease as another "state of the living body," which while distinct from health, was not in strict opposition to life as such.[15] The neuropathological correlate to Gaub's medical theories is found in the writings of Scottish physician and professor Robert Whytt (who lectured from Gaub's textbook), particularly his 1765 *Observations on the Nature, Causes, and Cure of those Disorders which Have Been Commonly Called Nervous, Hypochondriac, or Hysteric*. For Whytt nervous diseases emerged out of the normal functions of the nervous system, representing an *exacerbation* of normal neurophysiology, not its occasional erroneousness (which was the tendency taken by Whytt's contemporaries, including William Battie in his *Treatise on Madness* from 1757). For Whytt the fact that we are almost entirely "nervous" physiologically meant that we are always liable to becoming "nervous" in a pathological sense.[16] By the turn of the nineteenth century this compatibility of normal and abnormal states of the brain was a sentiment that Gall felt his composite brain could better depict, since as far as he was concerned, only his organological views could properly affirm how the brain could be "healthy and morbid, in a normal, and an anormal state, at one and the same instant."[17]

This transformation in the relationship between health and disease is a historical progression that I am, needless to say, perhaps, drawing from Georges Canguilhem. Canguilhem describes the tendency, increasingly apparent across the nineteenth century, to view pathology as an extension—a difference in degree—rather than a disparity or contradiction of normal physiology. For Canguilhem this sort of relationship between normality and pathology will be rigorously typified only in the writings of twentieth-century German neurologist Kurt Goldstein.[18] Earlier nineteenth-century variations tended to maintain, according to Canguilhem, a normatively idealized view of health, where pathology was still seen as a kind of "residue" of the norm.[19]

Be that as it may, this inter-embedded relationship between normality and pathology, characteristic of nineteenth-century medicine, will nonetheless be integral for the purposes of this analysis because the neuropathological disorder that perhaps most exemplified this relationship was also the disorder that was most formative to the work of John Hughlings Jackson—namely *epilepsy*.

## Normalized Abnormalities: The Case of Epilepsy

EPILEPSY WAS, HISTORICALLY speaking, a remarkable disorder. It was integral in the development of clinical neurology, being one of the most commonly treated neurological diseases of the period.[20] The National Hospital in London, where Hughlings Jackson at age twenty-seven was first employed as an assistant physician just two years after it opened in 1860, was devoted to "the relief of paralysis, epilepsy and allied diseases."[21] But epilepsy was also an illness that remained an object of psychiatric unease (and, as I describe later, medico-legal anxiety) until the turn of the century. It was, in many ways, a robust medical problem.

Epilepsy was pivotal for Hughlings Jackson's more mature, clinically derived theories of neurophysiology; it was the pathology on the basis of which he was able to reverse-engineer, as it were, a conception of healthy neurophysiology. It was his senior colleague, Charles-Édouard Brown-Séquard, who proposed that Hughlings Jackson consider disease as cases of the departure from normal neural processes.[22] Brown-Séquard had himself defined epilepsy as "an increased reflex excitability of certain parts of the cerebro-spinal axis, and in a loss of the control that, in normal conditions, the will possesses over the reflex faculty."[23] Hughlings Jackson's definition displayed some similarities: "Epilepsy is the name for occasional, sudden, excessive, rapid, and local discharge of grey matter,"[24] or what he elsewhere calls the "paroxysmal discharges" of "the cerebral hemisphere."[25] It was, in other words, an *unhealthy* release of nervous excitability, abnormal not in its occurrence but in its abruptness and degree of excess.

Reaching such a definition, however, involved abandoning prior methods of clinical pathology, which depended on the use of postmortem autopsies to correlate localized areas of apparent histological damage in the brain with recorded pathological symptoms.[26] Abnormal neural "discharges" were symptomatically displayed as convulsive fits, the seizures common to epileptic episodes. Hughlings Jackson proposed that the convulsive movements were a manifestation of neurological (specifically cerebral) damage, but not insofar as areas of the brain corresponded to individual muscles or parts of the body but rather because they represented *movements*. Given the way seizures "marched" up and down the body, usually beginning in a concentrated location and becoming increasingly generalized, epileptic convulsions effectively revealed to Hughlings Jackson the fact that the nervous system represented movements and motoric possibilities organized according to a hierarchy of complexity and specification. As he writes, "Nervous centers—we consider them as motor only—do not represent muscles but, or except as, movements. Each term or unit of a center represents a movement of the *whole* region which the center altogether represents." Epileptic attacks effectively demonstrated that the entire nervous system was a hierarchically arranged "sensori-motor machine."[27] "Speaking very roughly indeed, we may say that from the separate, detailed representation of parts of the body by the lowest centers up to the highest cerebral centers there is a gradual 'mixing up' by stages of increasingly complex representation, so that at the acme of representation each unit of the highest centers represents (re-represents) the whole organism in most complex ways, no two units of those centers representing all parts of the body in exactly the same degree and order."[28] By assembling and adopting this evolutionary framework of the nervous system, Hughlings Jackson was able to provide a compelling explanation of the phenomenon of an epileptic attack. That evolutionary framework allowed him to propose that "epilepsy [is] a disease of the highest cerebral centers. . . . The epileptic process begins in some part of the highest level (in some part of the 'mental centers') and when the fit is a severe one all the levels are greatly involved."[29] Any debilitation of the "highest level" of the nervous system would translate into a specific and contained tremor—the normal somatic commencement of an epileptic episode. (The mental or sensorial "auras" that often precede epileptic attacks were also signs that the highest levels had been debilitated first.) The transition to a more distributed convulsive fit meant that the epileptic discharge had extended farther down the neural hierarchy, affecting and aberrantly stimulating more levels and thus more generalized movements and motoric possibilities.

But by describing epilepsy in such a way, Hughlings Jackson believed he had effectively outlined the general processes of neurological disorders as such. In addition to its clinical-diagnostic sense, he deployed *epilepsy* in a more general, theoretical sense, as an umbrella term that could function as a template for neurological insanity.[30] Debilitations resulting from histological diseases, epileptic discharges, physical injuries, or even drunkenness were functionally analogous; they were all instances of what "epilepsy" in its general sense displayed: the dysfunction of a higher center and the consequential disinhibition of a lower level, either temporarily or permanently. Hughlings Jackson explains: "I take epileptic paroxysms and their after-conditions as being, together, an illustration of the morbid nervous affection of greatest complexity of symptoms. . . . The symptomatology of paroxysm is probably a universal symptomatology. . . . The study of epilepsy, therefore, involves the study of some cases of insanity. These insanities are what I have called after-conditions of the epileptic paroxysm."[31] Although epilepsy was not identical to the "after-conditions" and insanities that would arise only as a consequence of the aberrant neural discharges, still, epilepsy in very general terms offered an illustration of the neurological processes underlying, almost like a principle, the onset of almost all abnormal mental states. As William James affirmed in his *Principles of Psychology*, "Dr. Hughlings Jackson's explanation of the epileptic seizure is acknowledged to be masterly. It involves principles exactly like those which I am bringing forward here. The 'loss of consciousness' in epilepsy is due to the most highly organized brain-processes being exhausted and thrown out of gear. The less organized (more instinctive) processes, ordinarily inhibited by the others, are then exalted, so that we get as a mere consequence of relief from the inhibition, the meaningless or maniacal action which so often follows the attack."[32] In this sense epilepsy elucidated not merely the nature of neurological disorders but also the structure of the nervous system as such. After all, pathological volatilities were not opposed to normal neurological functions but grew out of them: "Symptoms of instability in disease are, I suppose, an exaggeration with caricature of the effects of healthy discharge."[33]

Epilepsy therefore could be viewed as a kind of blueprint outlining the brain's general sensorimotor physiopathology: "An ordinary severe epileptic 'attack' . . . is nothing more than a sudden excessive and temporary contention of very many of the patient's familiar normal movements—those of smiling, masticating, articulating, singing, manipulating, etc."[34] It was not that healthy movements were *un*-epileptic; they were simply not *yet* epileptic, not yet of such a degree that they could properly be called epilepsy. Hughlings Jackson explains that even "a sneeze is a sort of healthy epilepsy."[35] Normal neurophysiology was an unexaggerated analogue of neuropathology. It was not the case that epilepsy *could* occur, but that, from a certain point of view, it was always occurring *in potentia*, even if it was not always taking a strictly pathological form.

Hughlings Jackson's colleague, the psychiatrist Henry Maudsley, addressed the mental state of epileptic patients in *Responsibility in Mental Disease* (1874). The ostensibly abnormal epileptic state of mind, according to Maudsley, was not nearly as removed from normal subjective experience as one might initially believe. "There is a condition," he wrote, "intermediate between sleeping and waking in which, before consciousness is fully restored, the ideas and hallucinations of a dream persist for a time; so that a man, even though awake, shall think he sees the images or hears the voices of his dream." It was a condition, he explains, that we are often, and quite normally, prone to experience—a condition, Maudsley does not fail to add, that had even been the object of philosophical discussion by Aristotle, Spinoza, and others. "I doubt not," he concludes, "that in this condition of brief transitory delirium the mental state is very much like that which sometimes occurs in epileptics immediately after a fit, when on reviving to consciousness they break out into delirium; only it is of a much shorter duration."[36]

We are all, in some remote measure, already epileptic, or at least we are always liable to become so, given that the epileptic process already mirrors our physiology and many of our normally abnormal subjective mental states. As I detail further below, the concern with epilepsy was not that it blurred the line between health and disease. Rather it demonstrated the uneasy possibility that apparently normal neural and psychological conditions could already be, imperceptibly perhaps, entirely pathological.

## A Dangerous Disorder

NEAR THE BEGINNING of an essay published in 1874 in the *West Riding Lunatic Asylum Medical Reports* Hughlings Jackson writes, "There are few diseases of more practical interest than epilepsy. . . . The insanity of epileptics is often of a kind which brings them in conflict with the law. We have not only to treat epileptic patients, but we have occasionally to declare whether an epileptic is or is not responsible for certain quasi-criminal actions. The epileptic is beset with troubles."[37] More than just a disease of the mind and brain, epilepsy was viewed as a dangerous and often violent disorder, in which patients, either during or after their epileptic fits, would become susceptible to acting angrily, violently, and on occasion homicidally.[38] It was rare for epilepsy to be discussed without mention of the medico-legal implications to which the disease seemed—from as early as the 1840s to as late as the 1890s—almost intrinsically bound.

It was understood that in their postictal state, epileptic patients would display mental delirium and confusion and also a very manic, violent fury. Esquirol had warned in *Des maladies mentales* (1838), "The fury of epileptics bursts forth after the attacks, rarely before, and is dangerous, blind, and in some sort, automatic. . . . Epilepsy changes the character, and disposes the unfortunate subject to it, to bickerings and freaks of violent anger."[39] Wilhelm Griesinger in *Die Pathologie und Therapie der psychischen Krankheiten* (1845) had also described the "blind fury and violence" that would often arise as a consequence of the epileptic attack.[40] Such furious and violent behavior could, and apparently would, translate into the commission of numerous unintentionally criminal acts, and this dominant psychiatric assessment of the dangers of epilepsy was appropriated by the other behavioral and human sciences throughout the 1860s and 1870s.

The violence, moreover, was viewed as entirely intrinsic to the epileptic constitution; while not every epileptic was said to manifest violent symptoms, the possibility of violence, particularly the sort that tended toward criminal behavior, was nevertheless part and parcel of the disease. Epilepsy was not, to be sure, the only case where criminality and insanity were linked in some essential way. Maudsley proposed quite generally, "There is a borderland between crime and insanity, near one boundary of which we meet with something of madness but more of sin, and near the other boundary of which something of sin but more of madness."[41] The field of criminal anthropology emerged out of this belief in the inherent equivalence between insanity and violent criminal behavior. Indeed in 1889 Cesare Lombroso added the category of "the epileptic criminal" to the fourth edition of *Criminal Man*, writing, "We only have to stretch the definition of epilepsy a bit to draw a comparison between the psychological state of an epileptic during a fit and of the born criminal during his entire life."[42] Moritz Benedikt in *Anatomical Studies upon Brains of Criminals* anticipated Lombroso's sentiments by almost a decade when in 1878 he identified "the epileptic" as the "next blood kin" of "the constitutional criminal."[43]

While accepting the potential violent behavior inherent in the epileptic constitution, it was more common among medical practitioners to pathologize these postictal acts in order to set up the medical justifications by which epileptic patients could be legally exculpated from responsibility, should their behavior result in the commission of a crime. In *De l'état mental des épileptiques* (1861) Jules Falret suggested organizing all the varied mental pathologies associated with epilepsy under the term "epileptic insanity." As a diagnostic category, epileptic insanity would account for the "frequent complications of epilepsy," typically reported as "attacks of furious mania that follow epileptic attacks or alternate with them, and which carry the patient to the most violent, often the most dangerous, acts."[44] For Maudsley the manic fury of epileptic insanity gave it its "violent and destructive character," making it "a most dangerous form of insanity."[45] For Falret, however, the classification of epileptic insanity could medically adjudicate from the outset the patient's irresponsibility should the furious mania turn criminal. For some medical practitioners, such as Hughlings Jackson, it was of moderate importance to address the potential criminality of epileptic insanity; invoking Falret, Hughlings Jackson writes, "Epileptic insanity is usually violent. . . . The violence may take the form of crime from purely accidental circumstances."[46] For others epilepsy presented risks that demanded a vigilant attentiveness; in *The Physiology and Pathology of Mind* (1867), Maudsley alerts the reader that "the most desperate examples of homicidal impulses are undoubtedly met with in connection with epilepsy."[47]

Yet what was most unsettling about epilepsy was that its dangerousness was not restricted to the fact that the epileptic patient was prone to violence. It was, and continues to be, common to divide epilepsy symptomatically into some variation of a major or minor attack, the so-called *grand mal* and *petit mal* episodes, although the edges between the two intensities of attack were not always clearly defined. (Hughlings Jackson believed "these are not so much varieties as degree of the same thing."[48]) While the *grand mal* represented a full and usually debilitating seizure that consequently left the patient almost entirely unconscious, the *petit mal* represented only a partial episode, which could induce either a shorter duration of unconsciousness or bouts of dizziness or mental confusion.[49] It was the partial epileptic attack that became the object of particular concern, given that it could be so limited in scope as to go virtually unnoticed. The violence that could nevertheless follow a partial attack would appear to emerge, like an eruption, entirely out of nowhere. Esquirol had first cautioned, "There are cases, in which attacks occur suddenly and without any premonition, particularly in constitutional epilepsy."[50]

This attribute of epilepsy linked it to a broader medico-legal concern related to the difficulty of identifying madness in cases where a person does not *outwardly* appear insane but is nevertheless truly suffering from (and compelled to criminal behavior as a consequence of) some mental disturbance beneath the surface, as it were. In a short essay from 1861 titled "The Antagonism of Law and Medicine in Insanity and Its Consequences," Thomas Laycock explains that it is quite difficult to confuse cases of "stark, staring madness" with sane, premeditated behavior. But, he counters, "it is not such instances which give rise to doubts legal or medical; these arise as to much more subtle and insidious forms of disease and try the acuteness of the most experienced."[51] Epilepsy was a concern because it could take just such a "subtle and insidious" form. William Gowers, Hughlings Jackson's junior colleague at the National Hospital, explained that over and against cases of *grand mal* epilepsy, "the slight attacks of epilepsy vary much in character," insofar as "their nature is often not recognized by the patients or their friends."[52]

Epilepsy's true danger was a consequence of a very particular form that a partial epileptic attack might take. "It is customary to speak of a *Masked Epilepsy*," wrote Bucknill and Tuke in their *Manual of Psychological Medicine*.[53] This notion of "masked epilepsy"—a classification whose origin was attributed either to Esquirol or Benedict Augustin Morel (in a variation called "larvated" epilepsy)[54]—meant that instead of immediately following a clearly perceivable convulsive fit, the violent behavior would actually *replace* (therefore mask) the seizure itself; alternatively, the violence might indeed follow a seizure, but a seizure that was simply too slight to be perceived.[55] In a case of masked epilepsy the violent behavior could not be associated with an apparent convulsive attack. By not appearing as a complication or aftereffect of a seizure, there would no longer be any visible link between the violence and the recognizable disease of epilepsy.[56]

Maudsley believed that masked epilepsy was "a transitory mania occurring in lieu of the usual convulsions": "Instead of the morbid action affecting the motor centers and issuing in a paroxysm of convulsions, it fixes upon the mind-centers and issues in a paroxysm of mania, which is, so to speak, an epilepsy of mind."[57] In masked epilepsy, since the insanity was the only apparent symptom, the epileptic attack *was* (or at least appeared to be) the violent behavior itself (without a seizure). This meant that the pathological episode would not look like epilepsy at all, but merely the deliberative performance of excessive violence. "The condition is not merely of clinical interest," writes Gowers, "but also of medico-legal importance, since the performances may be complex, and may have all the aspects of deliberate volition; the initial epileptic seizure may be unnoticed by those around, and even unknown to the patient."[58]

The masked, "hidden," or "larvated" form of epilepsy was precarious at numerous levels. Because there was no forewarning (namely a seizure), a bystander could never be prepared for the potential eruption of manic fury. There were severe consequences for the epileptic patient as well, particularly when the question of responsibility was posed—since, as Gowers explains, the masked epileptic patient exhibits "all the aspects of deliberate volition," even though she would have been unconscious and would have retained no memory of her actions. The danger of epilepsy in its masked form was no longer embodied in the mere threat of manic and violent fury but in a potential commission of violence that would *appear* to have no epileptic symptomatology whatsoever— what Foucault calls "the zero degree of insanity,"[59] or a pathological violence, whose pathology was precisely the *absence* of pathological signs.

The true *problem* of masked epilepsy, then, hinged on the question of what, during the attack, was perceivable versus what remained imperceptible. Maudsley warned that the very existence of epilepsy could be overlooked "even by medical men" because in a masked state the epilepsy would appear as little more than "giddiness or faintness."[60] In *Responsibility in Mental Disease*, Maudsley relies on the work of an American psychiatrist, Manuel Echeverria, who in his 1873 essay, "Violence and Unconscious State of Epileptics, in Their Relations to Medical Jurisprudence," similarly warned that the epileptic attacks can display symptoms that remained "hidden and imperceptible, not offering the least warning to the patient, and unrecognized even by careful watchers."[61]

The neurological intervention and reformation of epilepsy, spearheaded by Hughlings Jackson, did not do away with the specter of its "masked" forms. Gowers insisted, "I believe that the old view [of masked epilepsy belonging to Esquirol and Morel] is not altogether untrue."[62] Instead of suggesting, as the psychiatrists had, that the violence of epileptic insanity *replaced* the fit, Hughlings Jackson merely argued that the fit usually existed but was often too slight to be noticed; epileptic insanities were almost always a postictal event.[63] Nevertheless the phenomenon of a "masked" epileptic attack persisted, and it continued to remain a problem of the discrepancy between what could and could not be perceived. In appearing entirely deliberative, volitional, and—for lack of a better word—*normal*, masked epileptic episodes did not provide the proper signals to indicate that the violence committed was the byproduct of illness. One could not outwardly judge what *should* be known about a person in a masked epileptic state—namely that there was *no person* there at all; that the patient was entirely unconscious, that her actions would not be remembered or accounted for. In masked epileptic states, then, a person's behavior could still appear deliberative and her actions would give no signs of illness. The normal behavior of personhood could, in illness, continue to be replicated; indeed that *was* the very pathology of the masked form.

### The Forensic Anxieties of Mental Automatism

BUT TO WHAT extent could any final determination be made about whether *someone* was or was not present and acting during the seemingly deliberative behaviors that arose in postictal states? This question represented another sort of anxiety with respect to epilepsy, specifically with respect to the behaviors exhibited during states of epileptic insanity especially in its "masked" form—a deeper anxiety that underlay but did not directly echo the alarm of the epileptic violence itself. I refer to the medical interest in what was seen to be the most peculiar aspect of epileptic phenomena: that the patient would transition into a state of *automatism* in which she would mechanically reproduce a variety of behaviors, many of which appeared perfectly normal. In his 1874 article published in the *West Riding Lunatic Asylum Medical Reports*, Hughlings Jackson emphasizes that it is important to consider a broader sample of epileptic behaviors, "not only cases of violent doings, but cases in which the patient simply acts oddly . . . cases in which there is no direct medico-legal interest." He continues, "The latter have nevertheless an important *indirect* medico-legal interest. It is convenient to have one name for all kinds of doings after epileptic fits, from slight vagaries up to homicidal actions. They have one common character—*they are automatic*; they are done unconsciously, and the agent is irresponsible. Hence I use the term *mental automatism*."[64] *Automatism* was the descriptive category utilized to describe the peculiar behaviors associated with epileptic insanity, whether those behaviors followed a discernible seizure or entirely replaced or "masked" the seizure.

As I'll describe in more detail, what made automatism such a strange phenomenon was that, as Hughlings Jackson explains, during states of automatic behavior, patients "may act in a very purpose-seeming way when unconscious,"[65] what Gowers also called the outwardly "deliberative" appearance of some "masked" postictal behavior. For Hughlings Jackson, while the threat of violence certainly demanded caution by the medical professional, it was the epileptic patient's tendency toward automatic yet apparently purposive behavior that presented a different kind of "indirect" medico-legal concern—a more fundamental forensic anxiety, I would propose, about the status of personhood during pathological states. Automatism was, after all, a behavioral condition at once apparently deliberative while, in theory, void of any actual deliberation—unconscious yet bearing the semblance of rationality.

By the middle of the nineteenth century automatism had emerged as an important concept in the psychophysiological study of both normal and abnormal behavior and states of mind. In England automatism had become initially promulgated in the work of Thomas Laycock and William Carpenter, both of whom proposed that there was a deep continuity, rather than decisive break, between higher mental processes and lower, automatic reflex actions of the nervous system.[66] This new approach introduced novel possibilities, but also new challenges, for the study of the brain and mind, particularly when the issue of automatism was considered through a set of popular debates instigated by T. H. Huxley's proposal in 1874 that all mental activity was, to some important degree, automatic. These debates were ultimately attempts to reconcile new psychophysiological paradigms with older, metaphysical commitments to the study of the mind.[67] Given that certain automatic behaviors were understood occasionally to surface, which in their appearance mimicked willful and thoughtful behavior (but which were not willful in themselves), it became important to determine whether classical properties of mental life (such as the will) should be eschewed or, instead, newly defended.[68] It was this impetus that motivated numerous medical and philosophical replies to Huxley, in particular by Carpenter and William James, but also a broader literary and cultural reaction to the fashion of automatism and to the new psychophysiological sciences more generally.[69]

In contrast to these more popular debates and philosophical discussions, automatism had become an integral diagnostic concept for clinical neurology and psychiatry. In 1874, a signal year for discussions related to automatism, Hughlings Jackson presented his own definition: "The mental automatism results, I consider, from over-action of the lower nervous centers, because the highest or controlling centers have been thus put out of use."[70] Drawing heavily on the work of Laycock, he effectively suggested that the exhibition of automatic behavior (accompanied by the absence of higher level mentation) was a primary symptom of the illness, disease, or injury of the brain.

It was not uncommon, in fact, for automaticity to be a synonym for pathology in general. In his 1878 Gulstonian lectures, David Ferrier links automatic states to the sorts of alienations suffered as a consequence of brain injury particularly to the prefrontal cortex. Invoking the famous case of Phineas Gage and other, similar reported cases, Ferrier proposes that as a consequence of certain neuropathologies, patients behave in ways that are "'purely automatic' or machine-like."[71] Maudsley insisted that in cases of childhood insanity, the child is transformed into a "little machine," either "mischievous" or "destructive" as the case may be.[72]

It was in relation to epilepsy, however, that automatism secured a robust clinical presence. Automatism was the condition that accounted for all the peculiar behaviors a patient was prone to experience and perform during bouts of epileptic insanity, including postictal mania as well as "masked" and seemingly deliberative actions. In his short medical treatise titled *Epilepsy* (1881), Gowers writes that automatic actions either follow or take the place of the epileptic attack and present a practical, medico-legal importance in excess of their clinical significance, in part because automatic actions "are sometimes complex and have the aspect of voluntary action." He continues, elaborating on concerns that had earlier been directed at the masked form of epilepsy, "It is, indeed, often not easy to convince observers that these actions are not deliberately volitional and intentional, so apparently conscious are patients; but consciousness is in an abnormal state, for the memory retains no recollection of these actions." The oddity of automatism was precisely the complexity it could manifest, the fact that a patient could automatically and yet unconsciously replicate many socially legible performances. In this sense automatism was linked to the more historically entrenched condition of somnambulism, "in which," Gowers reminds us, "the precision of muscular action is well known."[73] During a somnambulistic state, writes Maudsley, a patient "executes complicated acts of some kind which he could hardly do, and certainly could not do better, if he were awake."[74]

In the automatic state the patient was thought to be unconsciously reactivating habituated sensorimotor processes, actions that had effectively become automated and reproducible in the absence of any voluntary impetus. To the extent that the highest mental processes of the brain would have been viewed as continuous with the reflex and automatic functions of the nervous system—to the point that, for Hughlings Jackson, the entire nervous system (from top to bottom) was nothing more than a "sensori-motor machine"—then in theory *any* activity could be automatically replicated. Hughlings Jackson explains that during an epileptic attack, the transition to automatism might occur during a behavior "which is largely automatic, as, for example, playing a well-practiced tune." In such cases the patient, despite the attack, continues to play even while entirely unconscious. This is because "the automatic action had, so to speak, possession of the mind, and consciousness was not concerned in it, before the paroxysms occurred."[75] Hughlings Jackson is invoking a case first mentioned by Armand Trousseau in his *Lectures on Clinical Medicine* (1868). Trousseau, whose writings (along with Falret's) Hughlings Jackson found most instructive, describes the case of a violinist suffering from epilepsy, who occasionally suffers from attacks while playing but who continues to play, perfectly in time, while entirely unconscious, "as if," writes Trousseau, "those movements were guided by memory" and will.[76]

These automatic recuperations of habituated actions, however, were not always strict repetitions since they often presented some degree of novelty.[77] Trousseau explains, "The epileptic may complete the movements he has begun, and even perform new ones with a certain degree of regularity"; epileptic patients are also capable of "[answering] when spoken to, although they are not conscious of their answers."[78] Echeverria describes the case of a man suffering from a state of epileptic insanity lasting several days. While in a state of unconscious automatism, during which time he appeared to others entirely conscious and volitional, he agreed to be a sailor on a boat heading to London, although he had no knowledge of sailing.[79] Indeed the very possibility of committing a violent act, especially when the patient was not normally prone to violence, was indicative of the novel possibilities, dangerous or otherwise, inherent in automatic performances.

As a consequence of its complexity, what was most disarming about automatism was that it did not always exhibit, on its own, any particular mark of abnormality. This was precisely the anxiety that had developed around masked epilepsy; masked epilepsy came to be viewed as the imperceptible transition to an extremely complex automatism. Hughlings Jackson proposed that the slighter, more contained, and more imperceptible the epileptic attack was, the more complex and more elaborate the automatism would be.[80] In these automatic conditions, argued Falret, the patient "appears to have returned to himself; he enters into conversation with people around him, he performs actions that appear under the control of his will; he seems, in a word, returned to his normal state."[81] Echoing Falret, Maudsley describes the many instances of postictal behavior where "the individual appears completely restored to himself, and speaks and acts as if he were so," and that "this normal or apparently normal state of reason, in which he answers questions, makes remarks, and does various acts, may last for hours or even days."[82] If automatism was, in effect, the potential replication of the normal state, and if there were instances in which epileptic attacks would not manifest in the form of a recognizable seizure, then the onset of epileptic insanity—that is, the genuine transition from normality to pathology—would not in all instances register at a behavioral level. The patient would appear normal throughout.

The deeper challenge of automatism, however, was not in how normal behaviors might or might not be deceptively replicated. Automatism demonstrated the peculiar possibility that certain attributes of personhood could be simultaneously absent and present. The clinical interest in automatic conditions in part represented the rising medical and forensic acceptability of a certain paradoxical approach to the descriptive classification of a self. Echeverria maintains, "In regard to the unconsciousness of epileptics while thus rationally acting in their paroxysms of cerebral epilepsy, it is far from being an exceptional or unique phenomenon of the kind." It was by no means absurd to consider a patient to be without one mental attribute—consciousness—while in apparent possession of another, that is, reason or a rational propensity. It was possible to attend to personhood in this way and, more than possible, quite necessary, since as Echeverria asks, "how are we to decide on the legal responsibility" for patients in such a peculiar state of mind (at once a state of absent-mindedness)?[83] Indeed automatism occupied its own exculpatory role in Victorian forensic psychiatry, separate from other insanity defenses.[84] But this medico-legal consequence of automatism, I would propose, was subtended by a more fundamental forensic conundrum that automatism introduced—namely that the attributes of personhood could be present despite the paradoxical absence of the person herself.

Yet this conception that a patient could be outwardly and only apparently rational and normal while inwardly unconscious was itself the result of a certain view of what can be called an epileptic state of mind. One of the major transitions in the study of epilepsy that appears after 1870 is the belief that the epileptic attack does not result in the total loss of consciousness—in other words, that the epileptic state of mind should not be defined according to a strict dichotomy of normal consciousness and abnormal unconsciousness. Esquirol emphasized this sort of dichotomous view in 1838: "The pathognomonic character of epilepsy, consists in convulsions, the entire suspension of sensibility, and loss of consciousness."[85] But nearly half a century later Gowers would affirm, "Loss of consciousness is the rule, but exceptions are often met with, in which there is merely obscuration of consciousness for a few seconds, and no absolute loss."[86] What epilepsy and the peculiar automatic behaviors of epileptic insanity seemed to demonstrate was that consciousness need not be defined as a singular and static phenomenon of the mind. Maudsley refers to automatic behaviors as representing "peculiar states of epileptic consciousness" or "a pathological state of consciousness." He claims that some writers have been "in the habit of describing these *anomalous states of consciousness* as states of unconsciousness, moved thereto probably by the metaphysical notion of consciousness as a definite invariable entity which must either be or not be; but this is obviously a misuse of words; and what it behooves us to learn from them is that *consciousness is not a constant quantity*, but a condition of mind subject to manifold variations of both degree and kind."[87] What emerged by the mid-1870s and into the 1880s, then, particularly in British psychiatry and neurology, was a much more porous view of consciousness. Instead of a static phenomenon of the mind, conscious states were often viewed according to an oscillating spectrum of mental activity. Such a view could better accommodate how attributes of personhood could be both present and absent at once, without requiring the clinician or physiologist to contend with whatever paradoxes might arise—medically, philosophically, or legally—in declaring a patient to be unconscious yet simultaneously rational. Instead of a single consciousness, a person was said to possess various conscious states, some normal and others anomalous. Alfred Binet will claim in *Alterations of Personality* (1892) that some consciousness "can exist even when psychological activity is very low."[88]

In this way there could be, at one and the same time, *both* loss *and* retention of some mentation, and it was therefore not necessary to propose, in strict terms, that a person was absent even while many of the attributes of personhood were entirely active. Automatism could represent the porousness or continuity between conscious and semiconscious states of mind. Pathological reductions of mental activity could persist alongside the retention of some basic mental state, which in itself would be normal to some degree. By the mid-1870s personhood could indeed continue to persist, at least in some measure, during pathological episodes, in the form of automatic and quasi-conscious conditions. Pathology was not the empty and deceptive replication of normal, deliberative behavior. There was indeed *some* person there.[89]

### From Personal Identity to the Doubling of the Self

WHAT SUBSTANTIATED, AT least in part, the willingness to view postepileptic—or in fact *any* pathological—automatism as the persistence of *some* state of personhood was the rise of research on altered states of mind, in particular the interest in cases of double consciousness, which begin to be prominently reported in the mid-1870s. North Atlantic research into the doubling of personality was spurred by the French physician Eugène Azam and his studies, beginning in 1875, of the patient Félida X. As Hacking describes, the period from 1875 until the early 1890s witnessed a proliferation of cases of doublings and multiplications of conscious and semiconscious states. This proliferation, however, represented the intensification of medical reports and analyses of doublings that, particularly in Britain, were being investigated more than a half-century earlier.[90] While prompted by the faddishness of psychical research and spiritualism, particularly in Britain and the United States, though also in France, the neurological and medical-psychological interest in altered states of consciousness after the 1860s—and the unsettling degree to which consciousness could at least seem to *normally* take on an anomalous state—represented a rehabilitative effort to view such states as legitimate medical conditions.

Within the context of this effort automatism could be viewed as a manifestation of some sort of doubling of mental states. In 1873 Hughlings Jackson described postepileptic reductions to states of automatism as instances of a "double condition," where a patient experiences the loss of *some* consciousness and the simultaneous *retention* of some limited and automatic mentation: "The patient in epileptic mania is by *some process* suddenly reduced to the double condition. There is loss of consciousness, and 'under this' mental automatism goes on."[91] Of course he does not suggest that automatism amounts to the manifestation of a "double" or altered personality. Furthermore, at this point in his research he is not yet referring to the automatic state of mind lying "under" consciousness as another, "new" person (which, incidentally, he will about twenty years later). What he is instead proposing in his writings from the mid-1870s is that a person in a pathological state can nevertheless be subjected to a kind of partitioning of mental processes. This kind of language was not far off from certain explanations of the authentic condition of double consciousness.

A good example is Crichton-Browne's 1862 essay, "Personal Identity, and Its Morbid Modifications."[92] "The belief in personal identity," Crichton-Browne writes, "may be regarded as one of a series of fundamental and necessary ideas, with which the faculties *must* be invariably occupied in all actual states of consciousness."[93] Remnants of the Lockean conception of the self still loomed heavily in this period of nineteenth-century behavioral medicine. Théodule Ribot, at the outset of *The Diseases of Personality* (1885), a treatise Hughlings Jackson described as "a very valuable book," affirms the psychological significance of the category of "person," a term by which "we understand generally the individual, as clearly conscious of itself, and acting accordingly."[94]

For Crichton-Browne personal identity was a durable and resilient attribute of mind and not always susceptible to pathological disruption. Indeed, as he proposes, "errors of identity in the ordinary forms of mental disease are rarer than some psychologists have supposed." Only in cases of double consciousness, however, do we find genuine instances of the alteration of personal identity, "for the individual is separated into two distinct beings": "Mental identity is separated or multiplied into two distinct parts, so that two identities reside in the same individual."[95]

Over and against what William James will claim three decades later, for Crichton-Browne the doubling of consciousness was not a consequence of the lapses or other failures of memory.[96] It was instead the doubling of the mental process itself—that is, a bifurcation of consciousness into two separate tracks of self-identity rather than a severing of a single consciousness from itself. Crichton-Brown describes the case of a nameless female patient who presented symptoms similar to the more famous case of "J.H.," whom he introduced into the medical mythology of double consciousness.[97] The female patient exhibited exactly this sort of mental bifurcation: "These two states were in no way bound together, the patient's disposition, capacities, and attainments, being different altogether in each of them, so that she seemed alternately to be two distinct persons."[98] She displayed what could be called a highly differentiated and markedly bifurcated "double condition" of mind.

A similar argument is made over two decades later by the American neurologist Silas Weir Mitchell, whose 1881 *Lectures on Diseases of the Nervous System, Especially in Women* was dedicated to Hughlings Jackson.[99] In 1888 Mitchell recounts the case of the American Mary Reynolds who, along with Félida X, make up the most well-known cases of double consciousness in the nineteenth century. The case of Mary Reynolds had been relayed several times from as early as 1816; based on these prior case studies and his own more recent investigation of testimonials, Mitchell writes, "I am able to corroborate facts and from these various sources to supply the following account of these *two persons in one body*—two distinct lives antipodal from every mental and moral point of view." Mitchell espouses a similar conception that the doubling of personality was a result of the bifurcation of the mental process itself. Mary Reynolds was effectively two people, as far as Mitchell was concerned; she lived two separate and complete lives: "Each state had its mental accumulations. The thoughts and feelings, the likes and dislikes, of the one state did not in any way influence or modify those of the other."[100]

Only a few years after Mitchell's treatment of the Mary Reynolds case, Alfred Binet returned to the question of automatism, from the standpoint of a more receptive approach to doubled states.[101] Binet wonders whether instances of unconscious automatism "are also unconscious in themselves, or whether it is not more probable that they belong to a second consciousness." Instead of calling automatic states *unconscious* Binet proposes the term *subconscious*, something that can better accommodate the more unusual, though restricted mental states—such as "somnambulistic consciousness"—that emerge in pathological occurrences. For Binet it would not be correct to claim that when a patient's normal, full consciousness is debilitated in pathological circumstances, there is therefore *no* consciousness remaining. Rather it would be more correct to claim that *other kinds* of conscious states are exhibited when the patient's goes awry. There is always *some* mentation, even if it is not *one's own*. Binet writes, "Different states of consciousness may exist separately and without confusion in the same person, giving rise to the simultaneous existence of several conscious states, and even in certain cases to several personalities."[102]

Remarkably these "different" conscious states, while strictly disinhibited in pathological circumstances, were nevertheless exposed to some limited degree in normal behavior, primarily in the form of the various automatisms that pervaded everyday life: "Each one of us may, if we watch ourselves with sufficient care, detect in ourselves a series of automatic actions, performed involuntarily and unconsciously. To talk, to sit down, to turn the page of a book—these are actions which we perform without thinking of them." For Binet these infusions of automaticity meant that normal behavior itself consisted of oscillations across different genres of conscious states or different levels of mentation: "The unconscious movements of normal individuals should be considered . . . as the effects of very slight mental duplication." While the emergence of a full, second personality represented illness in a very significant and troubling sense, the occasional doubling of mental processes was not itself abnormal: "The rudiment of those states of double consciousness . . . may with a little attention be found in normal subjects," which for Binet corroborates "the formation of a center of consciousness functioning independently of the common center."[103]

If automatism displayed the behavioral attributes of personhood to such an extent that it could even develop into an alternate personality, then it would be valid to describe it as a conscious state—at the very least a limited and anomalous *sub*conscious state. Personhood did not need to be defined according to the consistency of a *single* state of consciousness since what everyday behavior in fact displayed was that it was normally proliferated by various automatic states of mind—that is, proliferated by a kind of multitude of alternate conscious modes. Personhood could embody not only the occasional inconstancy of consciousness but its peculiar abundance. While sometimes a person was *less* than herself, she was often actually *more*.

## Hughlings Jackson's "New Person"

ON THIS POINT I return to Hughlings Jackson and to his claims about personhood and pathology with which I began—namely that pathology corresponds to the emergence of a "new person." I have attempted to provide a sketch of discussions taking place in mid- to late nineteenth-century behavioral medicine that were concerned with how pathologically induced altered states of mind posed new medical, forensic, and epistemological challenges for the notion of personhood. I focused on a particular disorder—epilepsy and, more specifically, epileptic insanity—because of how formative that disorder was to the development of Hughlings Jackson's physiopathology of the brain. As I described earlier, what he saw in epilepsy and its convulsive symptomatology was a manifestation of the evolutionary structure of the nervous system. Postictal episodes of epileptic insanity (masked, violent, or otherwise) had a similarly demonstrative role to play. During episodes of epileptic insanity a patient was reduced to a state of automatism—which, I proposed, elicited as much epistemological anxiety about the status of personhood as it did medico-legal anxiety about the possible commission of violent crimes. For Hughlings Jackson epileptic insanity revealed that a person's pathological state of mind was composed of a "double condition."

Hughlings Jackson did not abandon this theory of the "double condition" or the duality of mentation after first proposing it in the early 1870s. Indeed more than a decade later it became the very paradigm he used to describe normal mental processes. In "An Address on the Psychology of Joking," delivered in 1887, he writes, "To borrow the ophthalmological term, we can say that mentation is "stereoscopic." . . . Just as there is visual diplopia so there is 'mental diplopia,' or, as it is commonly called, 'double consciousness.'" "Mental diplopia" occurs when present consciousness is interrupted by or superimposed on a prior or lower conscious state in the form of an association or a memory, which, Jackson emphasizes, is a remarkably common occurrence. "A smell," he writes, "say, of roses, I now have makes me think of a room where I passed much of my time when a child." The diplopia here occurs because the current scent of roses brings about "what we call the same smell, but really another smell," that is, the memory of a former scent. "The two scents," he explains, "linked together, hold together two dissimilar mental states (1) present, now narrowed, surrounding, and (2) certain vague quasi-former surroundings."[104] There is, in other words, in cases of "mental diplopia" a simultaneous reduction of present consciousness combined with the rise of another quasi-conscious mental state.

The example that Hughlings Jackson provides is intentionally ordinary; it is an effort to assert that everyday experiences are always shot through with this duality or "stereoscopy" of mentation. He presents another "exceedingly common" example of mental diplopia: "To further insist on the fact that mentation is stereoscopic, with more or less manifest diplopia, I give an example of mentation which is exceedingly common. Whilst writing I suddenly think of York Minster. Here is mental diplopia—(1) narrowed consciousness of my present surroundings, and (2) cropping-up of consciousness of some quasi-former surroundings. Of course something, whether I can mentally seize it or not, in my present surroundings, has developed a similar something associating with York surroundings." These bifurcations of thought are not instances of a single stream of consciousness undergoing associative leaps and breaks. They are instead the intermingling of two separate and distinct mental processes. As Hughlings Jackson asserts, "The process of all thought is double."[105] It is common for present consciousness to be prompted or occasioned to diminish, so that another, former (and in that sense "lower") consciousness can "crop up."

This phenomenon Hughlings Jackson calls "the 'play' of mind," and it is the means by which we are capable of generating incongruous mental states. Through these mental incongruities the mind performs the labor of synthesizing resemblances and differences between past and present experiences. This play of the mind, in other words, is the very basis for the more noteworthy products of intellection, including humor, creativity, and superstitious or magical reasoning—that is, mental performances that rely upon incongruities, contradictions, and even paradox.

But this *normal* stereoscopy of the mind is essentially analogous, in a much less exaggerated way, to what occurs in pathological states, where we observe a very similar kind of loss of higher consciousness simultaneously combined with disinhibition of a lower state of mind. Hughlings Jackson insists that "all morbid mental states are departures from normal mental states in particular ways—that, for example, the process of mentation in the maniac is but a caricature of that in healthy people."[106] Pathology is the manifestation of "morbid mental diplopia," which is only an amplification of its normal form. Mental automatisms, hallucinations, and "dream states" experienced during episodes of postictal epileptic insanity are all caricatures of the normal ways in which, within the confines of subjective experience, what is absent and past can often haunt the present. They are all iterations of the narrowing of present consciousness and the "cropping up" of another state of mind. As Hughlings Jackson explains in "On Neurological Fragments" (1892), "What we call the 'disorderly' mental condition of an insane man . . . is a mentation having the same laws as the mentation of [the] patient's former sane, or, as we may say, entire, self."[107]

Pathological iterations of morbid mental diplopia, however, did exhibit one very important difference. The "double condition" characteristic of pathology was not conducive to the sort of play of the mind occasioned in normal mental diplopia, where loss and retention was more or less balanced. In most pathological circumstances, present consciousness—which in normal states was only narrowed—was either entirely arrested or very nearly lost. The pathological stereoscopy was an imbalanced one, wherein the "narrowing" of present consciousness was more excessive than usual. Just how excessive that pathological narrowing could be actually varied, and that variation corresponded to the degree of mental impairment that a patient could suffer.

When abnormal narrowing was minimal, the morbid diplopia was also minimal—"shallower," as it were—and "there [was] but a slight departure from healthy action."[108] According to Hughlings Jackson, cases of slight morbid diplopia correspond to behavior that appears drunken or "dreamy"; indeed actual intoxication and dreaming while asleep were, as far as he was concerned, neurological correlates of the "dreamy" and drunken behaviors that arise as a consequence of authentic neuropathology.[109] "Insanity" for Hughlings Jackson was an extraordinarily broad category that diagnostically referred to *any* case of abnormal diplopia, for the morbid phenomenon underlying intoxication and dreaming was essentially analogous to postictal epileptic insanity (and other disorders), no matter the cause.[110]

While minimal pathological diplopia translated into only a slight departure from healthy behavior, a more excessive diplopia would translate into a more dramatic "reduction" in or "dissolution" of mentation, in which we observe the complex automatisms or the manic and violent fury characteristic of epileptic insanity. There were, in effect, different degrees of insanity, the intensity of which was based on the shallowness or depth of the abnormal diplopic reduction. It was this very point that Hughlings Jackson draws out in his 1894 article, "The Factors of Insanities," the text I described at the start of this paper. Degrees of insanity correspond to the intensity of diplopic reduction, which is itself a pathological dissolution of physiological function in the highest cerebral centers. As I mentioned at the outset, pathology corresponds to "dissolution" of cerebral processes, representing a reversal of the brain's normal and healthy evolutionary structure.

In "The Factors of Insanities," Hughlings Jackson explains how we can think of the degrees of insanity as they progress from shallower to deeper reductions. Every degree of insanity corresponds to an increasingly deeper diplopic reduction, where the patient suffers the double condition of "(1) negatively, defect of consciousness (loss of some consciousness) and . . . (2) the consciousness remaining."[111] More than that, each degree of insanity also corresponds to pathological dissolution of cerebral activity, a loss of function somewhere in the evolutionary hierarchy of the highest cerebral centers. Like the *mental* diplopia, the dissolution of a cerebral center still leaves intact the functioning of the subsequently lower (evolutionarily speaking) center. Hughlings Jackson schematizes these degrees of insanity in a graphic representation of four arbitrarily selected vertical layers marked A through D (fig. 4.1).[112] Each layer, he explains, stands in for a

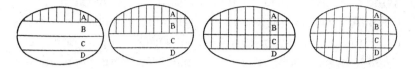

**Figure 4.1**   Diagrams of nervous dissolution, from Hughlings Jackson's "The Factors of Insanities" (1894).

level of mentation as well as a level of neural processing going on in the highest cerebral centers. Most remarkably, however, each layer also represents "different persons who have undergone that dissolution."[113]

Hughlings Jackson writes, "The only thing *disease* in the proper sense of pathological processes or result of one, has done, is to render functionless or to destroy the layer A, and we have necessarily to take into account the intactness, the functionability of layers B, C, and D." He graphically represents the first degree of dissolution, or the loss of A, as "–A+B+C+D." In this resultant state of brain activity and mentation, the shallowest degree of insanity, the patient has become, as far as Hughlings Jackson is concerned, an entirely *new person*: "In fact the whole person is now B+C+D. The term–A is only given to indicate how the new person, the man insane, B+C+D, differs from the former person, the sane man A+B+C+D." As a result of the abnormally diplopic reduction of mentation and the pathological dissolution of cerebral centers, a modification in personhood has taken place. "And here," explains Hughlings Jackson, "when using the term 'new person' . . . we must urge that what we call his delusions are his beliefs, and most generally his positive mental 'symptoms' are samples of a mentation which is only abnormal in contrast with the mentation of the prior self."[114]

Every subsequent degree of insanity corresponds to what Hughlings Jackson calls a "new person," and this continues until mentation has entirely ceased and all the highest cerebral centers have been dissolved; at this point "there is no *person*, but only a living creature."[115] "The Factors of Insanities" is not the first time Hughlings Jackson describes the relationship between personhood and pathology in this way. In "On Post-Epileptic States," (1888–89), he writes, "The insane man is a different person from his sane self, and we should take him up for investigation as that new person. . . . The insane man, the new person, has . . . a lower consciousness and shallower nervous system than the former person, his sane self."[116]

I would emphasize that Hughlings Jackson is not being figurative when he describes pathological episodes as the emergence of a "new person." After all, he was quite emphatic that an altogether different set of epistemological possibilities emerges for a patient suffering from pathological dissolution. The "neurological philosopher" that he was, Hughlings Jackson did not shy away from direct epistemological discussions.[117] As he explains in "Remarks on the Evolution and Dissolution of the Nervous System" (1887), all the data of experience—that is, our perceptions, ideas, memories, volitions, emotions, and so on—can be categorized as various states of what he calls an "object" consciousness. These states of object consciousness, however, are subtended by what he calls "subject" consciousness, an underlying foundation of mental activity. The dichotomy of object and subject consciousness represents a more fundamental "duality of consciousness," different even from the diplopia and stereoscopy that he otherwise describes. While subject consciousness corresponds to "consciousness of self," as Hughlings Jackson explains, "self is not describable psychically in the same way that object consciousness is." Subject consciousness cannot, like object consciousness, be the focus of direct experience or comprehension. "Subject consciousness," he writes, "is something deeper than knowledge; it is that by which knowledge is possible. . . . It is an awareness of our existences as individuals, as persons having the objective states making up for each, the (his) Universe."[118]

Subject consciousness is something akin to the preconditions of personhood, the underlying condition of mind that makes personhood possible as such but doesn't determine any of the particularities of an individual person. Object consciousness, on the other hand, fulfills that demand by filling in the experiential data that make us *this or that* person. Pathological reductions in mentation occur at the level of object consciousness, resulting in changes in the data of experience, in what a person remembers, perceives, wants, or believes.[119] What remains intact, however, even in instances of insanity is subject consciousness. Pathological reductions do not necessarily disrupt the *possibility* of personhood itself, but only the specific makeup of the person; they disrupt states of object consciousness, thereby altering a patient's specific mental geography, making her someone else. Only when the possibility of experiencing has ended, that is, only when mentation is absolutely dissolved and the highest cerebral centers entirely lost (when "there is no *person*, but only a living creature"), do we say that subject consciousness was dissolved as well.

### Making Up a Neural Person

THUS FOR HUGHLINGS Jackson pathology represented a modification of neural and mental processes to such an extent that it became justifiable to classify pathological subjectivity as the emergence of an altogether new person. Personhood, then, was allowed to remain a persistent attribute of both normal and abnormal neural and mental activity precisely because the suffering patient became *someone else* during her illness rather than no one at all.

There is, as I have proposed throughout this essay, an important historical justification that substantiates the authenticity underlying Hughlings Jackson's proposal that a "new person" properly emerges in pathological occurrences. It was becoming increasingly legitimate in behavioral medicine, I have argued, to view automatism as a certain limited state of personhood; directly linked to this was the increasing willingness to consider and the increasing acceptability of considering the possibility that mentation could bifurcate within one individual into two separate persons. Hughlings Jackson's claims, then, are very clearly intelligible within and animated by the background of this medical reasoning, one in which "insane *person*" was no longer a contradiction in terms. The larger discourse surrounding the pathology of personhood in behavioral medicine comprised a "space of possibilities" through which Hughlings Jackson's own specific claims exhibit a historical viability.

Hughlings Jackson, however, did not rely on the various medical rationalizations circulating at the time, forgoing any explicit invocation of the discourse of "doubles," alters, or multiple personalities. While his work is firmly situated within the behavioral medicine of altered states of mind, he developed his own neuropsychological paradigm through which the relationship between personhood and pathology could be expressed. Instead of replicating the broader discussions of the pathology of personhood, he recalibrated them and grounded them on a novel conceptual framework—one that was, coincidentally, revolutionary in the modernization of neurology and the physiopathology of the brain.

There is, then, an important yet inadvert consequence to Hughlings Jackson's contributions to the pathology of personhood, which is that the category of person and the category of neural function became in his writings dramatically formalized in relation to each other. More noteworthy than the fact that personhood was altered during pathological occurrences is the fact that personhood was now, thanks to Hughlings Jackson, neurologically guaranteed *even during* pathological occurrences.

Into the broader discourse of behavioral medicine between 1860 and 1900, a discourse that sought to present a viable relationship between madness and self, Hughlings Jackson introduced the value of neurological reasoning. It was a reasoning that not only consolidated and stabilized the relationship between insanity and the person but also transformed personhood and its various modifications into an object that was best attended to by the neurological study of mental states. Neurology implicitly established itself as a system of knowledge that could effectively unify all the pathological disunities of the self: where there is brain, there will always (some) person be. Here is perhaps one of the earliest and most robust formalizations of "neural person-hood," a certain way of expressing the truth of the self, where neurological discourse accounted for the multifaceted dimensions of personhood and for all its possible and unsettling dispersions as well. Yet such a possibility did not arise because Hughlings Jackson felt it was necessary or valuable to conflate or confuse people with their brains. It arose, inadvertently perhaps, through his efforts to address a set of problems in behavioral medicine, problems related to how personal identity was being modified in light of research directed at altered states of mind.

# Endnotes

1. John Locke, *An Essay concerning Human Understanding* (Oxford: Oxford University Press, 1975), book 2, chapter 27, pp. 328–48. "Of identity and diversity" is an expansion and elaboration of a briefer discussion of personal identity from book 2, chapter 1.

2. Ibid., 346. Etienne Balibar writes, "This unity that we have called (since Kant) the subject, and that Locke—the first to do so—calls self-consciousness, is indissolubly logical (self-identical), moral and legal (responsibility, appropriation) and psychological (interiority and temporality). It secretly calls itself *My Self.*" *Identity and Difference: John Locke and the Invention of Consciousness*, trans. Warren Montag (New York: Verso, 2013), 61.

3. Locke, *Essay*, 342, emphasis added.

4. William James also describes a pathologically debilitated brain as "virtually a new machine." *Principles of Psychology*, 2 vols. (New York: Henry Holt, 1890), 1: 143.

5. Franz Gall, *On the Functions of the Brain and Each of its Parts*, 6 vols., trans. Winslow Lewis (Boston: Marsh, Capen & Lyon, 1835), 3: 76.

6. I discuss this relationship between a unified mind and homogeneous white matter in "The Brain and the Unconscious Soul in Eighteenth-Century Nervous Physiology: Robert Whytt's *Sensorium Commune*," *Journal of the History of Ideas* 74, no. 3 (2013): 425–48.

7. James Crichton-Browne, "Personal Identity and Its Morbid Modifications," *Journal of Mental Science* 8 (1862–63): 385–95, 535–45. See also Janet Oppenheim, *"Shattered Nerves": Doctors, Patients, and Depression in Victorian England* (New York: Oxford University Press, 1991), chapter 2.

8. Ian Hacking, "Historical Ontology," in *Historical Ontology* (Cambridge, MA: Harvard University Press, 2002), 23.

9. Ian Hacking, "Making Up People," in *Historical Ontology*.

10. Thomas Willis, *Five Treatises: 1. Of Urines, 2. Of the accension of the blood, 3. Of musculary motion, 4. The anatomy of the brain, 5. The description and uses of the nerves* (London: T. Dring, C. Harper, J. Leigh, and S. Martin, 1681), 91; Humphrey Ridley, *The anatomy of the brain* (London: Printed for Sam. Smith and Benj. Walford, 1695), 91.

11. René Descartes, "Letter to Princess Elizabeth, 8 July 1644," in *The Philosophical Writings of Descartes*, vol. 3, ed. John Cottingham, Robert Stoothoff, Dugald Murdoch, and Anthony Kenny (Cambridge: Cambridge University Press, 1991), 237, emphasis added.

12. Descartes, "Letter to the Marquess of Newcastle, October 1645," in *Philosophical Writings of Descartes*, 275.

13. Descartes, "Letter to Christophe Villiers, 30 July 1640," in *Philosophical Writings of Descartes*, 149. For more on Descartes's baroque politics, see

Victoria Kahn, "Happy Tears: Baroque Politics in Descartes *Passions de l'âme*," in *Politics and the Passions, 1500–1850*, ed. Victoria Kahn, Neil Saccamano, and Daniela Coli (Princeton, NJ: Princeton University Press, 2006). I discuss absolutist motifs in seventeenth-century neuroanatomy in "Material Translations in the Cartesian Brain," *Studies in History and Philosophy of Biological and Biomedical Sciences* 43 (2012): 244–55.

14. Thomas Hobbes, *Leviathan* (Cambridge: Cambridge University Press, 1991), 230.

15. Jerome Gaub (H. D. Gaubius), *The Institutions of Medical Pathology*, trans. Charles Erskine (Edinburgh, 1778), 9. See also Thomas H. Broman, "The Medical Sciences," in *The Cambridge History of Sciences*, vol. 4: *Eighteenth-Century Science*, ed. Roy Porter (Cambridge: Cambridge University Press, 2003), 478.

16. I provide a detailed analysis of Whytt and the embedded relationship between normal and pathological nervous processes in "The Brain and the Unconscious Soul."

17. Gall, *On the Functions of the Brain*, 2: 300.

18. Georges Canguilhem, *The Normal and the Pathological* (New York: Zone, 1991), especially 181–201. See also Georges Canguilhem, "The Normal and the Pathological," in *Knowledge of Life*, trans. Stefano Geroulanos and Daniela Ginsburg (New York: Fordham University Press, 2008).

19. Canguilhem, *The Normal and the Pathological*, 85–86.

20. W. F. Bynum, "The Nervous Patient in Eighteenth- and Nineteenth-Century Britain: The Psychiatric Origins of British Neurology," in *The Anatomy of Madness: Essays in the History of Psychiatry*, vol. 1: *People and Ideas*, ed. W. F. Bynum, Roy Porter, and Michael Shepard (New York: Tavistock, 1985), 96.

21. Gordon Holmes, *The National Hospital, Queen Square, 1860–1948* (London: E&S Livingstone, 1954).

22. Kenneth Dewhurst, *Hughlings Jackson on Psychiatry* (Oxford: Sanford, 1982), 13.

23. Charles Édouard Brown-Séquard, *Course of Lectures on the Physiology and Pathology of the Central Nervous System: Delivered at the Royal College of Surgeons of England in May, 1858* (Philadelphia: J. B. Lippincott, 1860), 183.

24. John Hughlings Jackson, "On the Anatomical, Physiological, and Pathological Investigation of Epilepsies," in *Selected Writings of John Hughlings Jackson*, 2 vols., ed. James Taylor (New York: Basic Books, 1958), 1: 100, quoted in George K. York and David A. Steinberg, *An Introduction to the Life and Work of John Hughlings Jackson with a Catalogue Raisonné of His Writings* (London: Wellcome, 2006), 13; also quoted in Owsei Temkin, *The Falling Sickness*, 2nd edition (Baltimore: Johns Hopkins University Press, 1971), 337.

[25] John Hughlings Jackson, "On the Scientific and Empirical Investigation of Epilepsies," in *Selected Writings*, 1: 177. See also John Hughlings Jackson, "Remarks on Dissolution of the Nervous System as Exemplified by Certain Post-Epileptic Conditions," in *Selected Writings*, 2: 9.

[26] For more on Hughlings Jackson's methodological approach, see York and Steinberg, *An Introduction*, 9–13.

[27] John Hughlings Jackson, "On Some Implications of Dissolution of the Nervous System," in *Selected Writings*, 2: 38, 29. Hughlings Jackson's views anticipated the experimental validation of the fact that cortical regions localized complex sensorimotor processes; they also affirmed a view first presented in the writings of Thomas Laycock that the highest cerebral processes were not different in kind from peripheral reflex action. See Kurt Danziger, "Mid-Nineteenth-Century British Psycho-physiology: A Neglected Chapter in the History of Psychology," in *The Problematic Science: Psychology in Nineteenth-Century Thought*, ed. William R. Woodward and Mitchell G. Ash (New York: Praeger, 1982). See also Edwin Clarke and L. S. Jacyna, *Nineteenth-Century Origins of Neuroscientific Concepts* (Berkeley: University of California Press, 1992), chapter 6.

[28] John Hughlings Jackson, "On Post-Epileptic States," in *Selected Writings*, 1: 372. Put another way, Hughlings Jackson writes, "The highest level consists of the highest motor centers (pre-frontal lobes) and the highest sensory centers (occipital lobes). They represent all of the body triply indirectly. These highest sensori-motor centers make of the 'organ of the mind' or the physical basis of consciousness; they are evolved out of the middle, as the middle are out of the lowest, and as the lowest are out of the periphery; thus the highest centers re-re-represent the body—that is, represent it triply indirectly." John Hughlings Jackson, "Remarks on the Evolution and Dissolution of the Nervous System," in *Selected Writings*, 2: 79.

[29] Hughlings Jackson, "On Post-Epileptic States," 1: 372.

[30] Temkin, *The Falling Sickness*, 339.

[31] John Hughlings Jackson, "Ophthalmology and Diseases of the Nervous System," in *Selected Writings*, 2: 347.

[32] James, *Principles of Psychology*, 2: 126. James, however, has apparently misinterpreted Hughlings Jackson's use of the term *organized*, assuming that it is synonymous with *complex*. The lowest centers are, for Hughlings Jackson, the *most* organized in the sense that they are the oldest, simplest, most automatic, and most stable. The highest centers, which are the newest and most complex, are the *least* organized and therefore the centers most liable to debilitation. Hughlings Jackson writes, "The doctrine of evolution implies the passage from the most organized to the least organized, or, in other terms, from the most general to the most special" (*Selected Writings*, 2: 58). See also Macdonald Critchley and Eileen A. Critchley, *John Hughlings Jackson: Father of*

*English Neurology* (New York: Oxford University Press, 1998), 56 and chapter 7 overall.

33 Hughlings Jackson, "On the Anatomical, Physiological, and Pathological Investigation of Epilepsies," 1: 96.

34 John Hughlings Jackson, "On Convulsive Seizures," in *Selected Writings*, 1: 431.

35 Hughlings Jackson, "On the Anatomical, Physiological, and Pathological Investigation of Epilepsies," 1: 96.

36 Henry Maudsley, *Responsibility in Mental Disease* (New York: D. Appleton, 1876), 251–52, 253.

37 John Hughlings Jackson, "On Temporary Mental Disorders after Epileptic Paroxysms," in *Selected Writings*, 1: 119.

38 Temkin, *The Falling Sickness*, 266–69.

39 J. E. D. Esquirol, *Mental Maladies*, trans. E. K. Hunt (Philadelphia: Lea and Blanchard), 150, 170.

40 Wilhelm Griesinger, *Mental Pathology and Therapeutics*, trans. C. Lockhart Robertson and James Rutherford (London: New Sydenham Society, 1867), 405.

41 Maudsley, *Responsibility in Mental Disease*, 34.

42 Cesare Lombroso, *Criminal Man*, trans. Mary Gibson and Nicole Hahn Rafter (Durham, NC: Duke University Press, 2006), 259.

43 Moritz Benedikt, *Anatomical Studies upon Brains of Criminals: A Contribution to Anthropology, Medicine, Jurisprudence, and Psychology*, trans. E. P. Fowler (New York: Wm. Wood, 1881), 158 (published in German in 1878). See Arnold I. Davidson, *The Emergence of Sexuality: Historical Epistemology and the Formation of Concepts* (Cambridge, MA: Harvard University Press, 2001), 219.

44 Jules Falret, *De l'état mental des épileptiques* (Paris: P. Asselin, 1861), 25, 2, my translation.

45 Maudsley, *Responsibility in Mental Disease*, 228.

46 Hughlings Jackson, "On Temporary Mental Disorders after Epileptic Paroxysms," 1: 120.

47 Henry Maudsley, *The Physiology and Pathology of the Mind* (New York: D. Appleton, 1867), 309.

48 Hughlings Jackson, "On Temporary Mental Disorders after Epileptic Paroxysms," 1: 121.

49 Temkin, *The Falling Sickness*, 257–60.

50 Esquirol, *Mental Maladies*, 147.

51 Thomas Laycock, *The Antagonism of Law and Medicine in Insanity and Its Consequences* (Edinburgh: Oliver & Boyd, 1862), 4.

52  William Gowers, *A Manual of Diseases of the Nervous System*, American edition (Philadelphia: Blakiston, 1888), 1090.

53  John Charles Bucknill and Daniel Hack Tuke, *A Manual of Psychological Medicine: Containing the lunacy laws, the nosology, etiology, statistics, description, diagnosis, pathology, and treatment of insanity* (London: J. & A. Churchill, 1874), 339.

54  Temkin, *The Falling Sickness*, 317–20.

55  Ibid., 317–18.

56  It was this fact that, for Lombroso, made epilepsy most akin to criminality: "The sharp, sudden outbursts that characterize hidden epilepsy demonstrate that this affliction is similar to criminality and moral insanity" (*Criminal Man*, 257).

57  Maudsley, *Responsibility in Mental Disease*, 230.

58  Gowers, *A Manual of Diseases of the Nervous System*, 1092.

59  Michel Foucault, "About the Concept of the 'Dangerous Individual' in Nineteenth-Century Legal Psychiatry," in *Power: Essential Works of Foucault, 1954–1984*, vol. 3, ed. James D. Faubion (New York: New Press, 1994), 181.

60  Maudsley, *Responsibility in Mental Disease*, 243–44.

61  Manuel Gonzales Echeverria, "Violence and Unconscious State of Epileptics, in Their Relations to Medical Jurisprudence," *Journal of Insanity* (April 1873): 526.

62  Gowers, *A Manual of Disease of the Nervous System*, 1092.

63  See specifically Hughlings Jackson, "On Temporary Mental Disorders after Epileptic Paroxysms," 1: 119–34. See also William Gowers, *Epilepsy, and Other Chronic Convulsive Diseases: Their Causes, Symptoms, and Treatment* (New York: William Wood, 1885), 100.

64  Hughlings Jackson, "On Temporary Mental Disorders after Epileptic Paroxysms," 1: 122.

65  Hughlings Jackson, "Remarks on Dissolution of the Nervous System," 2: 10.

66  Danziger, "Mid-Nineteenth-Century British Psycho-physiology," 124–33.

67  Lorraine J. Daston, "The Theory of Will versus the Science of Mind," in Woodward and Ash, *The Problematic Science*, 100–103; Thomas Henry Huxley, "On the Hypothesis That Animals Are Automata and Its History," *Fortnightly Review* 22 (1874): 555–80.

68  Jonathan Miller, "Going Unconscious," in *Hidden Histories of Science*, ed. Robert B. Silvers (New York Review, 1995), 20; Alison Winter, *Mesmerism: Powers of Mind in Victorian Britain* (Chicago: University of Chicago Press, 1998), 287.

69  W. B. Carpenter, "On the Doctrine of Human Automatism," *Contemporary Review* 25 (December 1874–May 1875): 397–416, 940–62; William James, "Are We Automata?," *Mind* 4, no. 13 (1879): 1–22. See Anne Stiles, *Popular Fiction and Brain Science in the Late Nineteenth Century* (Cambridge: Cambridge University Press, 2012), 50–56; Vanessa L. Ryan, *Thinking without Thinking in the Victorian Novel* (Baltimore: Johns Hopkins University Press, 2012).

70  Hughlings Jackson, "On Temporary Mental Disorders after Epileptic Paroxysms," 1: 123.

71  David Ferrier, *The Localisation of Cerebral Disease Being the Gulstonian Lectures of the Royal College of Physicians for 1878* (London: Smith, Elder, 1878), 35. Ferrier is citing a case reported by "Dr. Davidson" in the *Lancet* a year earlier (1877).

72  Maudsley, *The Physiology and Pathology of the Mind*, 276, 283.

73  Gowers, *Epilepsy*, 95–96, 97. The full quote reads, "The state [of automatism] is no doubt closely allied to the condition of somnambulism, in which the precision of muscular action is well known."

74  Maudsley, *The Physiology and Pathology of Mind*, 267.

75  Hughlings Jackson, "On Temporary Mental Disorders after Epileptic Paroxysms," 1: 124.

76  Armand Trousseau, *Lectures on Clinical Medicine, Delivered at the Hôtel-Dieu, Paris*, 5 vols., trans. P. Victor Bazir (London: New Sydenham Society, 1868), 1: 59.

77  See Hughlings Jackson, "On Temporary Mental Disorders after Epileptic Paroxysms," 1: 124.

78  Trousseau, *Lectures*, 1: 58, 60.

79  Echeverria, "Violence and Unconscious State of Epileptics," 545.

80  Hughlings Jackson, "On Temporary Mental Disorders after Epileptic Paroxysms," 1: 123.

81  Falret, *De l'état mental des épileptiques*, 73–74.

82  Maudsley, *Responsibility in Mental Disease*, 237.

83  Echeverria, "Violence and Unconscious State of Epileptics," 552, 549.

84  Joel Peter Eigen, "Delusion's Odyssey: Charting the Course of Victorian Forensic Psychiatry," *International Journal of Law and Psychiatry* 27 (2007): 395–412; Joel Peter Eigen, *Unconscious Crime: Mental Absence and Criminal Responsibility in Victorian London* (Baltimore: Johns Hopkins University Press, 2003), chapter 6, "Crimes of an Automaton."

85  Esquirol, *Mental Maladies*, 151.

86  Gowers, *A Manual of Diseases of the Nervous System*, 1091.

87  Maudsley, *Responsibility in Mental Disease*, 238–39, emphasis added. See Danziger, "Mid-Nineteenth-Century British Psycho-physiology," 136.

88  Alfred Binet, *Alterations of Personality*, trans. Helen Green Baldwin (New York: D. Appleton, 1896), 64.

89  This was, of course, not a universal position. There were some who continued until the end of the century to advocate for a strict distinction between the pathology of unconscious automatism and what would otherwise be a healthy state of mind. Pierre Janet is a good example of such a view. In demonstrating a commitment to a unified conscious self (which was an institutional commitment to Cousinian philosophy), Janet continued throughout the 1880s and 1890s to view automatism as an impoverishment of unified consciousness, a psychological "depersonalization." See, for example, Jan Goldstein, "Neutralizing Freud: The Lycée Philosophy Class and the Problem of the Reception of Freud in France," *Critical Inquiry* 40, no. 1 (2013): 76.

90  Ian Hacking, *Rewriting the Soul: Multiple Personality and the Sciences of Memory* (Princeton, NJ: Princeton University Press, 1995), chapter 11. See also Ian Hacking, "Double Consciousness in Britain, 1815–1875," *Dissociation* 4, no. 3 (1991): 134–46; Ian Hacking, "Two Souls in One Body," *Critical Inquiry* 17, no. 5 (1991): 838–67.

91  John Hughlings Jackson, "Remarks on the Double Condition of Loss of Consciousness and Mental Automatism Following Certain Epileptic Seizures," *Medical Times and Gazette*, July 19, 1873, 63.

92  Hacking argues that this essay represents the culmination of explicit British discussions of the doubling of personality ("Double Consciousness in Britain," 142). I would propose that it instead marks the transformation to a more diffuse medical discussion of the meaning and possibility of mental "doublings" and bifurcations.

93  Crichton-Browne, "Personal Identity," 388.

94  Dewhurst, *Hughlings Jackson on Psychiatry*, 116; Théodule Ribot, *The Disease of the Personality* (Chicago: Open Court, 1895), 1.

95  Crichton-Browne, "Personal Identity," 391, 395, 536.

96  Ibid., 536. See James, *Principles of Psychology*, 1: 379.

97  Hacking, "Double Consciousness in Britain," 142.

98  Crichton-Browne, "Personal Identity," 539.

99  S. Weir Mitchell, *Lectures on Diseases of the Nervous System, Especially in Women*, 2nd edition (Philadelphia: Lea Brothers, 1885). See also Dewhurst, *Hughlings Jackson on Psychiatry*, 121.

100  S. Weir Mitchell, *Mary Reynolds: A Case of Double Consciousness*, reprinted from the Transactions of the College of Physicians of Philadelphia, April 4, 1888 (Philadelphia: Wm. J. Dornan, Printer, 1889), 2, 14, my emphasis.

James reviews Mitchell's treatment of Reynolds in *Principles of Psychology*, 1: 381–84.

[101] Hacking, *Rewriting the Soul*, 97.

[102] Binet, *Alterations of Personality*, 98, 153, 155.

[103] Ibid., 97, 221, 83.

[104] John Hughlings Jackson, "An Address on the Psychology of Joking," in *Selected Writings*, 2: 359, 361.

[105] Ibid., 2: 362.

[106] Ibid.

[107] John Hughlings Jackson, *Neurological Fragments*, ed. James Taylor (London: Oxford University Press, 1925), 48.

[108] Hughlings Jackson, "Remarks on the Double Condition," 63.

[109] Hughlings Jackson writes, "For an hour or two after injuries to the head there may be a condition very like that of a man slightly drunk" ("On Neurological Fragments," in *Neurological Fragments*, 52); "In both dreaming and drunkenness the state of things is closely analogous" ("Remarks on the Double Condition," 63).

[110] Hughlings Jackson explains, "We require for the science of insanity a rational generalization, which shall show how insanities in the widest sense of the word, including not only cases specially described by alienists, but delirium in acute non-cerebral disease, degrees of drunkenness, and even sleep with dreaming are related to one another" ("Remarks on Dissolution of the Nervous System," 2: 4–5). See also Dewhurst, *Hughlings Jackson on Psychiatry*, 73.

[111] John Hughlings Jackson, "The Factors of Insanities," in *Selected Writings*, 2: 414.

[112] These diagrams appear in the original publication of Hughlings Jackson's article.

[113] Hughlings Jackson, "The Factors of Insanities," 2: 413.

[114] Ibid., 2: 414, 416.

[115] Ibid., 2: 417.

[116] Hughlings Jackson, "On Post-Epileptic States," 1: 383.

[117] Temkin, *The Falling Sickness*, 354.

[118] Hughlings Jackson, "Remarks on Evolution and Dissolution of the Nervous System," 2: 96. See also Anne Harrington, *Medicine, Mind, and the Double Brain* (Princeton, NJ: Princeton University Press, 1987), 226–30.

[119] Hughlings Jackson, "Remarks on Evolution and Dissolution of the Nervous System," 2: 91.

*Stefanos Geroulanos and Todd Meyers*

# 5 Integrations, Vigilance, Catastrophe: The Neuropsychiatry of Aphasia in Henry Head and Kurt Goldstein

THIS ESSAY FORMS part of a broader project concerned with the ways in which, around World War I, the disciplines dealing with the human body exhibited a marked shift toward medical and physiological theories of bodily integration.[1] The rationale for this shift resided in the idea that this integration could be observed and studied at times when the body was threatened with or was undergoing collapse. The body was not only compensating for injury and loss but was also capable of bringing itself to collapse in its effort to regulate itself and respond to external aggression. War neuroses, including those caused by brain injury, often seemed too complex to classify, their symptoms often too variegated to fit a single formal description set in clear primary and secondary symptom categories.[2] As a result patients suffering from war neuroses needed to be treated as individuals, *operationalized as individuals* so as to negotiate the often new or compound symptomatologies and pathologies that had emerged, and their particularity or individuality needed to be communicated and aggregated for some sense of breadth and particular modalities of care to be established.

This was the case as well with forms of somatic (neurophysiological) integration resulting from or undone by traumatic brain injury. This essay focuses on the work of Henry Head and Kurt Goldstein, two neurologists often read in tandem as aphasiologists who became influential through their work with brain-injured soldiers during and after World War I. As soldiers with traumatic head injuries poured into hospitals, both Goldstein and Head, who had already studied aphasia for some time, found that "the war was producing a series of cases unique in the history of the subject."[3] Goldstein would later claim that, while only 4 to 10 percent of brain-injured soldiers survived in the 1914–17 period, he had nevertheless carried out a "systematic" study of two thousand patients and highly detailed research on some "90 of them regularly over the following decade."[4] Working independently Head and Goldstein advanced a forceful critique of the existing understanding of aphasia and linked disorders; their predecessors had by and large sought to explain the loss of certain capacities to use language on the basis of the cerebral localization of linguistic centers, and damage to particular areas was understood to cause damage to particular ways of using language. But to them, as to contemporaries ranging from August Bethe to Karl Lashley,[5] this was not just nonsense at the diagnostic and theoretical levels; it guaranteed that the medical and psychiatric experience of the patient would be one marked and signed through and through by violence. Head famously charged aphasiologists from the turn of the century with being "compelled to lop and twist their cases to fit the procrustean bed of their hypothetical conceptions."[6] Both, moreover, went much further than merely offering a critique of an aphasiology that was conceptually lacking or experimentally dated in its obsession with cerebral localization, its limited understanding of how language works, and its obsolescence in relation to neurological theory. They pursued a sense of how the healthy mind operates, how it is integrated, how individuals function *as individuals*, how they partake of language as a whole, how they respond to injury and loss.

When we note that this essay zooms in on a kind of organismic integration and collapse, we mean several things by *integration*: first, the integration of the nervous system, which extended to an integration of the many mechanical elements of the body into a single whole; second, the integration settled and understood, both *a contrario* and experientially, as the instance of health right before and around the collapse caused or precipitated by brain injury. But we argue that despite remarkable similarities, Head and Goldstein produced quite radically different conceptions of the organism in that they also worked with a third concept of integration and disintegration, as each of them saw its effects very differently. This was for them a philosophical as much as a therapeutic issue.

We first touch on Charles Scott Sherrington's understanding of integration in the nervous system, which lay at the foundation of neurological and neurophysiological conceptions of integration since the publication of his *Integrative Action of the Nervous System* in 1906.[7] Sherrington plowed the land on which neurology would situate the problem of bodily integrity, fundamentally displacing earlier attempts to think wholeness at the corporeal level. We then discuss the development of Head's and Goldstein's ideas, particularly the structure of their theories of wholeness. In the final section we focus on the divergence between Head's *Aphasia and Kindred Disorders of Speech* and Goldstein's *The Organism*, attending to the profound differences resulting from the different conceptual and experimental priorities of the two thinkers, as well as the value and significance of disintegration as they imagine it.

## Sherrington's Integration

HEAD AND GOLDSTEIN were by no means alone in proposing that the brain or the nervous system be understood as a single whole, even as the way they argued this point was largely their own. Often contrasted with the more traditionally mechanistic approach that composed the organism of its constituent parts or attended only to certain parts, especially in experimental psychology, the claims for the integration of the self and the brain had become widely available in the prewar period and included evolutionary approaches in late nineteenth- and early twentieth-century biology, the holism of Gestalt psychology, and, no less significant for our purposes here, that of John Hughlings Jackson's "Spencerian" approach to aphasia.[8] In his Croonian Lectures of 1884, "The Evolution and Dissolution of the Nervous System," Hughlings Jackson utilized a Spencerian framework to establish a sense of the nervous system evolving from an automatic, animal basis common to all organisms that share it to a developed, highly specialized, voluntarist system in humans; injured human beings would find their highly developed system regress to lesser status.

The definitive work that rendered integration into a significant component of neurology's research program was Charles Scott Sherrington's pioneering *Integrative Action of the Nervous System*, which was published in 1906. Of Sherrington's theory we need to retain two elements here. First, nervous integration is central to the body, its capacity to move and conduct and respond to stimuli as a whole, even in order to stand still. It occurs first at the level of the reflex arc, that is, of "the unit mechanism of the nervous system when that system is regarded in its integrative function. *The unit reaction in nervous integration is the reflex*, because every reflex is an integrative reaction and no nervous action short of a reflex is a complete set of integration. The nervous synthesis of an individual from what without it were a mere aggregation of commensal organs resolves itself into co-ordination by reflex action." As reflexes involve the most basic coordination of sensation and motor response to a stimulus, they provide something like the alphabet out of which animal organisms are phrased. Sherrington goes on to postulate the "purely abstract conception" of the *simple reflex*, a "convenient if not a probable fiction" according to which each single reflex action could be simply distinguished from the rest of the activity of the nervous system as it unites the organism.[9] From this most basic integrative unit, the reflex arc, Sherrington proceeds to ever more complex units (the segment and the segmental series), discussing integration at local and regional levels.

Thus, second, the brain and the central nervous system more generally become the central, "dominant" site for this integration. The central nervous system is "an organ of co-ordination in which from a concourse of multitudinous excitations there result orderly acts, reactions adapted to the needs of the organism, and . . . these reactions occur in arrangements (patterns) marked by absence of confusion, and proceed in sequences likewise free from confusion. By the development of these powers the synaptic system with its central organ is adapted to more speedy, wide, and delicate co-ordinations than the diffuse nervous system allows. Out of this potentiality for organizing complex integration there is evolved in the synaptic nervous system a functional grading of its reflex arcs and centres." Sherrington's nervous integration concerns the body and the body alone; there is no sense in these passages that at stake is the organism's behavior in relation to its outside world nor that anything other than direct neurophysiological experimentation can be used to engage these questions: "The motile and consolidated individual is driven, guided, and controlled by, above all organs, its cerebrum. The integrating power of the nervous system has in fact in the higher animal, more than in the lower, constructed from a mere collection of organs and segments a functional unity, an individual of more perfected solidarity."[10] The kinds of integration that we will attend to in the cases of Head and Goldstein are, as we shall see, very much in critical conversation with Sherrington's, with marked differences in tone, object, and argumentation.

### Head and Goldstein: Parallel Introduction

HENRY HEAD IS best remembered as a major figure in early twentieth-century English neurology, though, as L. Stephen Jacyna has argued in a detailed recent biography, his care was ultimately "medical science" and the ways it provided for patients, and he was almost as much a man of letters as of neurons.[11] Following nearly forty years of work in neurology and neuropsychiatry, Head proposed in his 1926 *Aphasia and Kindred Disorders of Speech* a general approach to disorders of speech, which expanded to a theory of nervous functioning, consciousness, and normal and pathological behavior that strove toward an understanding of the ways behavior broke down in the brain-injured organism and was abnormally reconstructed by it. Kurt Goldstein's career dates to the 1900s; what defined him as a neurologist and a psychiatrist, however, was his work at the Frankfurt Neurological Institute during and after the war, which led to publications of therapeutic and experimental clinical case studies of brain-injured soldiers (many of them coauthored with the psychologist Adhémar Gelb and sometimes with other colleagues from the Neurological Institute) and the development of an approach to therapeutics. Goldstein wrote books and a plethora of papers on the treatment, retraining, and rehabilitation of brain-injured soldiers as well as on forms and effects of aphasia.[12] *The Organism* (*Der Aufbau des Organismus*, 1934) constituted the second intellectual event of Goldstein's career; in it he went so far as to propose a general theory of the organism specifically intended as a broad and radical critique of neurological and physiological theories that, to him, underappreciated and misunderstood its integrated structure. Goldstein focused on the way an organism compensates for lost performances and the way the loss of an autonomous, abstract level of thinking as a result of brain injury results in complex and body-wide counterefforts.

Head and Goldstein did not know each other personally, and they do not appear to have maintained a correspondence. Still Head concluded his 140-page history of the understanding of aphasia in *Aphasia and Kindred Disorders of Speech* with a highly favorable discussion of Gelb and Goldstein's analyses of the Schneider case and "A Case of Color Amnesia," describing them as "admirable work along psychological lines" and "a notable exception" to existing approaches.[13]

In the single letter from Goldstein surviving in Head's archive, Goldstein noted his appreciation for the high regard in which Head held his work at a time when Goldstein and Gelb's work still lacked broad recognition in Germany, and he complimented Head's book as "a standard work in our research."[14] For his part Goldstein was well aware of Head's work, advocating its value in his discussions and correspondence with his cousin Ernst Cassirer.[15] The convergence of their intent offers a quite astounding sense of scientific practitioners working in parallel;[16] even granting a certain reliance of each researcher on the other's work in the 1920s, it is striking to see the parallels of their paths, their critiques of aphasia and brain localization, and their efforts to understand somatic and nervous integration.

Both Head and Goldstein trained in neurology in the 1880s and 1890s. For Head, who grew up and began his studies in Cambridge, England, the defining years took place in Halle, Germany.[17] Goldstein trained under Carl Wernicke in Breslau. Both arrived at the problem of aphasia before the war—by Head's own account by 1910, and judging by Goldstein's publications by 1906, when he was still at Königsberg.[18] In 1914 Goldstein moved to Frankfurt as director of the Institute for Research into the Effects of Brain Lesions in the Frankfurt Neurological Institute,[19] becoming, like Head in London (first at the London Hospital and as of 1915 at Empire Hospital), a major figure in the study of brain injury, aphasia, and the treatment and rehabilitation of war neuroses involving actual cerebral injury. Each moreover carried out a significant part of his definitive research with psychologists: Goldstein's principal collaborator over two decades was the Gestalt psychologist Adhémar Gelb, who remained his principal collaborator until Gelb's death, during the time that Goldstein was writing *The Organism*. In Head's case this was the psychologist W. H. R. Rivers, who, together with his own erstwhile disciple Charles S. Myers, had participated in the Torres Straits anthropological expedition of 1898 and who in 1903 had carried out with Head the famous experiment in nerve regeneration, where, after Head's

**Figure 5.1**  Henry Head, his eyes closed, facing away, while W. H. R. Rivers
uses a long pin or pencil to trace the spaces of protopathic and epicritic
sensibility on Head's arm. Courtesy of Wellcome Library, PP/HEA B.9.
Source: Wellcome Library, London. Sir Henry Head (1861–1940), Archives
and Manuscripts. CMAC PP/HEA/B.9. "W.H.R. Rivers and Sir Henry
Head experimenting on nerve function. circa 1905." Copyrighted work
available under Creative Commons Attribution only licence CC BY 4.0.

radial nerve was surgically spliced, the two of them carried out
sensation tests on his arm over a period of two years (fig. 5.1).[20]

By 1918 both Head and Goldstein felt uneasy in their discipline:
each sought to emphasize the credentials of their work across a
broader field of study than neurology alone, and each expressed
and pursued a philosophical and methodological bent.

Rather than delimit exact points of influence, it is worth emphasizing parallels that will eventually help us see the considerable difference between Head's and Goldstein's conceptions of the self.[21]
Both sought to explode, in different but parallel ways, the diagnostic and explanatory category of aphasia, decrying existing formulations as profoundly insufficient and criticizing (in Head's case

quite harshly) their predecessors. They shared a number of targets, including brain localization, that is, the tendency to identify faculties (e.g., language, sensation, memory) with particular sites in the brain, which was a traditional concern of psychology and one that had become central to anatomy and physiology since Franz Joseph Gall. Both of them denied that such faculties existed at all in a manner separable from the organization and integrations of the nervous system that underwrote and substantiated consciousness—and both studied the very problem of this consciousness, its relation to language as much as to its neuronal scaffolding, at length. As a result both disclaimed the prevalent psychological and physiological expectations of most of their predecessors and contemporaries to be insufficient to the point of harmful to the patients under their care, patients whose highly individual reaction to brain injury and aphasia was routinely whitewashed.[22] Both found considerable inspiration in the writings on language and aphasia by Hughlings Jackson, whom they treated as prefiguring their own work, a *vox clamantis in deserto* amid the nineteenth-century obsession with localization.[23] At the same time, both Head and Goldstein declined to use the evolutionary language that Hughlings Jackson (following Herbert Spencer) and Sherrington relied on; their engagement with the integration of the nervous system and with aphasia would be based on pathological and neurophysiological mechanisms, not predicated upon Darwinian evolution. Finally both described aphasia as a particular disorder of broad consequence for the organism. For Head, at stake was not only the behavior of an organism limited in its access to language, nor just the improvement of retraining and rehabilitation practices; at stake was the very relation between neurological integration and the use of language as a fundamental whole constitutive of social experience. For Goldstein it was the subsumption of the nervous system into the capacity of the healthy organism to abstract from current "concrete" circumstances, to maintain an organismic wholeness, a "categorical attitude," and a decisive measure of autonomy vis-à-vis his or her environment, to "self-actualize." Both came to uphold aphasia as *the* pathological condition whose study facilitated an understanding of the structure of the self. Head's and Goldstein's subsequent paths involved generalization from brain injury to organismic and subjective wholeness. For Head the concern with the brain's access to language—the ontology that makes language possible and constant—would become central; Goldstein would turn instead to organismic norms and health. The similarities end at this point. Working in parallel, at a distance, they proceeded to establish the brain-injured organism as a unit at the verge of collapse *in very different terms*. What follows is a consideration of each of their approaches to aphasia and their way of handling integration. These were philosophical, therapeutic, and also technical issues, a matter of how tests, patients, and revisionism contributed to two very specific senses of integration and collapse.

### Henry Head's Orders and Disorder

HEAD'S PRINCIPAL PUBLICATIONS on aphasia span the period from 1915, when he republished Hughlings Jackson's essays on aphasia in a double issue of the journal *Brain*, to 1926, when he assembled his two-volume magnum opus, *Aphasia and Kindred Disorders of Speech*, out of considerably expanded versions of his publications from 1919 to 1924 and a set of twenty-six case studies, each of them detailed to between nine and forty pages in length. By 1915 Head had already drafted his critique of contemporary aphasiology, writing in his introduction to this republication of Hughlings Jackson's essays, "It is generally conceded that the views on aphasia and analogous disturbances of speech found in the text-books of to-day are of little help in understanding an actual case of disease."[24] Hughlings Jackson's work, which "prefigured" this "more than fifty years ago,"[25] deserved to be treated *as current* and new because his "psychological" approach undermined precisely these textbook reductions and confusion.

Narratively—for it is a work acutely aware of its own rhetorical force—*Aphasia and Kindred Disorders of Speech* presents us with a movement from harsh epistemological critique to climactic resolution. "I have attempted to blaze a track through the jungle," Head writes,[26] and he means several jungles, which he superimposes on one another: the jungle created by theories of aphasia, which he considers profoundly misdirected and by this point jumbled into mutually contradictory positions; the jungle experienced by each individual patient, thanks to the specific collapse of the order of language each has to endure; and the jungle of conceptual and clinical obstacles plaguing the physician's clear understanding of the individuality and condition of the patient. The book is structured to open out these fields one by one.

The first volume aims to balance extensive historical, theoretical, and clinical claims. Head opens not with a précis of his project but with two histories: a highly abbreviated autobiographical account of his engagement with the problem of aphasia and a 150-page history of the theory and treatment of aphasia since the early nineteenth century. His history "cannot be an unprejudiced account";[27] it is directed to support his argument and to extract and redeploy useful concepts, such as Constantin von Monakow's *Diaschisis* (splitting). Head is particularly harsh toward the progress of the field since Pierre Marie, titling this chapter "Chaos" and featuring only Gelb and Goldstein's as a serious current theory, indeed as the last word in a specifically psychological attempt to resolve the violence forced on the patient throughout this history.[28] Except for himself and them, aphasia's history is ostensibly the persistence of a theory that must finally be undermined, a theory that has relied on the identification of lesions in the brain with particular linguistic capacities, and that has been largely the consequence, he argues, of the absorption into neurology of elements from its theoretical and the broader philosophical milieu: "New wine is poured into old bottles with disastrous results."[29]

Head then offers a chapter on experimental and clinical methodology and follows it with a clinical chapter referencing in moderate detail five of the cases he would engage later.[30] The litany of his objections to existing aphasiology is strewn across these pages, as if the battle against traditional interpretations had to be fought over and over, "a restless destruction of the false god," while the promised catharsis is repeatedly deferred.[31] There exist no "centers" for the use of language in any form; we do not think in images,[32] nor in words as imagined in the nineteenth century. We cannot distinguish a priori between conditions of sensory or motor aphasia or between categories like ataxia, agnosia, amnesia, aphasia, which are "descriptive terms only" and not "isolated affections of speaking, reading and writing."[33] At the same time, and based on this clinical material and revisionist theory, Head begins the ascent of his argument toward its central thesis and resolution.

Aphasia should no longer be understood as a disorder of the brain and mind, subdivided into different, well-organized, well-known, localizable varieties, but rather as a group of *disorders of symbolic thought and expression*.[34] As Marcel Mauss, among others, would learn from Head,[35] these disorders never affect merely one element of language and symbolic use and do not correlate with specific injuries: different elements (naming, syntax, speaking, comprehension, and "internal speech") are all affected, but not to the same degree, with the result that what is necessary is a schema for understanding how injury affects particular patients and groupings of patients. Aphasic disorders are not "static" conditions but dynamic "disorders of function," and they are best understood, generally speaking, in terms that correlate *some* brain injury with a "want of capacity to use language with freedom." In this, different integrations play a remarkable role: "We cannot even analyse normal speech strictly in terms of motion or sensation; most acts depend on both these factors for perfect execution. For the use of language is based on integrated functions, standing higher in the neural hierarchy than motion or sensation, and, when it is disturbed, the clinical manifestations appear in terms of these complex psychical processes; they cannot be classed under any physiological categories, motor or sensory, nor even under such headings as visual and auditory." While it is essential to produce new categories, or "headings," to group and generally distinguish among different groups of errors and difficulties affecting the patient, "these designations are empirical; they have no value as definitions and must not be employed to limit the extent of that loss of function to which they are assigned."[36]

Let's reconsider this set of critiques by way of the affirmative (as opposed to the critical) argument Head proposes regarding aphasia and integration. Tucked halfway through the book is the perhaps definitive claim, the center out of which Head will fight toward his narrative and experimental climax: "Every case of aphasia is the response of an individual to some want of power to employ language and represents a personal reaction to mechanical difficulties in speech. No two patients exhibit the same signs and symptoms; still less do the phenomena of recovery follow exactly the same course. But by comparing on broad lines the manner in which function is restored in cases of each particular variety of aphasia, it is obvious that some acts of speech are recovered sooner than others, and we thereby obtain an insight into the essential nature of the clinical phenomena."[37] Head's explicit goal is the restoration of lost functions (and, alongside this, the patient's recovery and the broader importance of the clinical data). But it is the first sentence that offers the striking formulation: every case is *the response of an individual*; it is *a personal reaction*. Head does not simply write about individual cases, the difference of the experience for each patient ("no two patients").[38] Instead he crafts individuality itself in terms of *each patient's dynamic response to loss, lack, or difficulty*. In the phrase above, the brain-injured patient's self can now be found in the organism's compensatory and transformative effort at recovery—in the reconstruction of a relationship to language that has been broken and, with it, a relationship to other people. The locus of the individual is not to be found in the brain, nor even in the healthy individual, but in the *now individualized* pathological and recuperative relation to symbolic formulation and expression, a relation that is *individualized because it is not "free."*

That the response should be individual is decisive for several reasons that deserve closer attention: it is a dynamic process with a social component, an integrative process, a process that concerns language far more than it concerns any component of man's biological substratum, a process that unveils for us the different levels of the somatic, conscious, and linguistic integrations that form selfhood.

First, aphasia is a functional and dynamic, as opposed to a static, condition: in the aftermath of a wound, the patient undergoes a continuing experience, not merely a stagnation, and this experience is at times one of recovery and at times one of further loss and decline, generally depending on the existence of any progressive disease of the brain and also practices of socialization.[39]

As a result, second, the actual site of injury is less significant than the integrative way the organism attempts to respond and compensate. Head is emphatic on this point concerning the site and totality of the new self. "In conclusion," he writes at the end of the book, "I have attempted to show that, when some act or process is disturbed in consequence of an organic or functional lesion, the abnormal response is a fresh integration carried out by all available portions of the central nervous system. It is a total reaction to the new situation. The form assumed by these manifestations cannot be foretold by a priori consideration, but must be determined by observation and described in terms of the affected process." Elsewhere, in an almost warlike formulation of the organism as an arsenal of powers, he writes, "Faced with a new situation, the organism puts out all its powers, conscious, subconscious and purely physiological, in order to produce an adequate response directed towards its welfare as a whole."[40] At stake, then, is the sense that the organism's survival and welfare depend on a new, recovered integration, on the way this integration compensates for the original, normal, now lost integration. The patient is not some stable being that is tied either to the period prior to his injury or to his injury. Instead he is "all the powers," the "abnormal response" that is this "fresh integration"—a being imagined in continuing resistance and response to the loss.

Third, as should be clear by now, Head understood aphasia as first of all a language disorder and concomitantly a psychological disorder (insofar as language has an "emotional" and not only a formal component). Only then is it also a neurological and biological disorder, insofar as the disruption of normal neuronal integration conditions and the organism's response to that disruption ground the loss biologically. He writes, "Evidently there must exist a group of functions indispensable for language in its widest sense," and even if these functions are not necessarily equally essential, they are "affected by the destruction of parts of the brain."[41] Thus a lesion involves the disturbance of cerebral and physiological processes and, as a result, suffering. Though he kept copious notes and diagrams of both the sites of injury and the kind and extent of neurological damage involved, Head does not seem to have found significant use for them. Instead he articulated a hermeneutic scheme for dealing with patients that was fundamentally concerned with patients' success or failure to sustain a linguistically articulated order, to understand the meaning of the words or sentences they were uttering, to copy text and images, to identify regular and frequently used "symbolic expressions" (e.g., the time on a clock). Patient individuality cannot therefore be approached through the physiological site of damage, but only through the linguistic one, and as a result diagnosis and treatment remained tied precisely to his language.

Fourth, the conception of language proposed by Head allows for concurrent processes of generalization, categorization, and precision. Head offers four headings—*verbal, syntactical, nominal,* and *semantic*—and emphatically characterizes them as rough classifications, as "designations" that are "empirical" and "have no value as definitions and must not be employed to limit the extent of that loss of function to which they are assigned."[42] Note, in the following formulations, how Head not only refuses localization and biological reduction but also sees and categorizes the aphasias in terms solely structured by way of access to language and communication.

(1) Verbal aphasia "consists mainly in a defective power of forming words, whether for external or internal use." (2) Syntactical aphasia, "this disorder of symbolic formulation and expression[,] consists essentially in lack of that perfect balance and rhythm necessary to make the sounds uttered by the speaker easily comprehensible to the audience." (3) Nominal aphasia, "this disorder of language[,] consists essentially of [an] inability to designate an object in words and to appreciate verbal meaning" (fig. 5.2). Finally, (4) semantic aphasia involves cases wherein "comprehension of the significance of words and phrases as a whole . . . is primarily disturbed" (fig. 5.3).[43]

Loss of capacity for language had of course been the definitive characteristic of aphasia since Broca, and categories like "word-blindness" and "word-deafness" had proliferated since Charlton Bastian.[44] Head, who had studied with Bastian, duly exhibits his customary ruthlessness toward his predecessors, especially the premise that language is merely a dimension of sensory or motor systems and that therefore aphasia is but a simple destruction of only this dimension. Establishing his own theory in tandem with the quadripartite argument regarding the disruptions of symbolic thought and expression, Head argues that "we do not think in words," by which he means, in explicitly neogrammarian fashion, that words are merely a small part of a language, that they have multiple meanings, that language must itself be understood as a dual system of its own,[45] at once logical and psychologi-

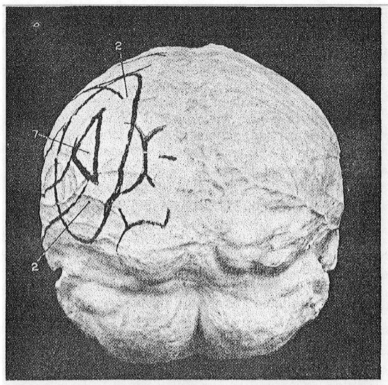

Fig. 11. To show the site of the injury to the brain, seen from behind, in No. 2 and No. 7, cases of Nominal Aphasia.

**Figure 5.2** Henry Head's own localization efforts reported in *Aphasia and Kindred Disorders of Speech* (1: 457). Cases of nominal aphasia. Note case 7, retraced here.

cal, and indeed as one that we normally engage as a whole. The aphasic, Head claims, is "like a man talking a foreign language": "Debarred by his disability from the use of verbal forms natural to him . . . [he] seeks some other and less difficult mode of conveying his meaning."[46]

This linguistic and psychological approach to aphasia is not, nevertheless, the final word—and despite supporting the "psychological" approaches of Goldstein and Hughlings Jackson, Head does not describe his own approach in those terms.[47] To complete

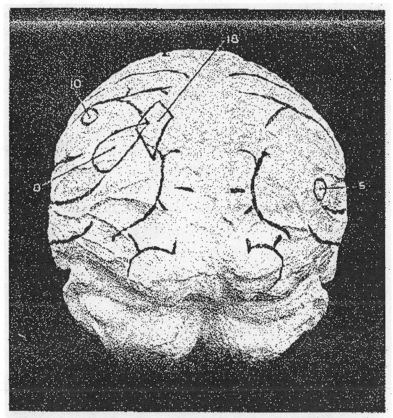

Fig. 13. To show the site of the injury to the brain, seen from behind, in No. 8, No. 10 and No. 18 to the left and No. 5 to the right of the middle line. All these patients suffered from Semantic Aphasia.

**Figure 5.3** Henry Head's own localization efforts reported in *Aphasia and Kindred Disorders of Speech*. Cases of semantic aphasia.

his own account and offer a neurological grounding for his theory, Head complements linguistic and pathological psychology and case-based psychology with a neurobiology of the levels of integration. Pathology shows disturbance in both internal interpretation and language participation. The central term in this account, and perhaps the central theoretical term in this last stage of his work, is *vigilance*. Sherrington's writings and experiments play a significant role in underwriting this concept, as Head attempts to explain how an activity that is apparently so fundamental to the organism's functioning as is language (or, more precisely, that broader activity he calls "symbolic thought and expression") can be so fundamentally disrupted due not to the loss of a part of the brain but to the loss of a certain integration.

To establish *vigilance* Head proceeds from neurophysiological examination, using the example of a human being whose spine has been severed alongside Sherrington's cat decortication experiments. Removing the cat's brain or handling a patient who has lost the use of his legs reveals for Head a fundamental propensity of the organism to carry out certain acts beneath any conscious level. Here is the theory of "protopathic" and "epicritic" sensibility that he and Rivers had developed in the famous earlier experiment on nerve regeneration. As sensation returns to the affected area, it is first the "deeper," far more intense, and undifferentiated sensation that returns, with the higher-end "epicritic" sensation returning and allowing for an integrated consciousness only later.[48]

*Vigilance* refers to integrated activity carried out at the barest neurological level. The word indicates the persistence of function during a moment or protracted period of apparent sleep, the capacity to keep vigil over the body. Head holds that the body itself, disencumbered of the will, indeed even in the absence of a cerebrum, keeps vigil over itself. It maintains a set of very basic postures and responds to certain forceful excitations depending on the circumstances (e.g., temperature or other elements of the milieu) and can even have certain localized reflexes, regardless of paralysis. Vigilance is "high-grade physiological efficiency," exhibiting "purposive adaptation" to external stimuli,[49] and giving the impression of carrying out conscious acts—acts we know to be impossible as acts but that the body itself manages to maintain. Vigilance thus becomes a key concept at this point: an integration that occurs at the purely neurophysiological level, that underwrites any epicritic sensation and conscious act, and that exceeds and brings together more local integrations that may occur at the neurological level. It allows Head to posit that, from muscle tone to even conscious actions, a series of integrations occur, of which the body is not aware. And it explains the loss of language in terms that finally include both lesion and language on the basis of effects on the general "neural potency" of the organism:

> Every automatic act is an exercise in physiological memory and can be disturbed by vital states which have nothing to do with consciousness. What wonder that the complex powers demanded by speech, reading, and writing, can be affected by a lesion, which diminishes neural vitality. Vigilance is lowered and the specific mental aptitudes die out as an electric lamp is extinguished when the voltage falls below the necessary level. The centres involved in those automatic processes, which form an essential part of the conscious act, may continue to live on at a lower vital level, as when under the influence of chloroform; they do not cease entirely to function, but the vigilance necessary for the performance of their high-grade activities has been abolished by the fall of neural potency.[50]

Head can then finally conclude the book in a manner that at once radicalizes this kind of holism, positing that each mind acts as a whole, that each "conscious act is a vital process directed toward a definite end and the consequences of its disintegration appear as abnormal modes of behavior."[51] *This can be understood only thanks to the event of "disintegration," which demonstrates in and thanks to pathology what has otherwise been normal for the organism, necessary for the continuous functioning of consciousness.* Consciousness does not normally recognize the limits of its wholeness: "The continuity of consciousness . . . is produced by habitually ignoring the gaps."[52] Only once it has been broken can the facts and function of consciousness be recognized. Indeed *not one but several integrations*, "a series of integrations" take place, "most of which take place on a purely physiological level," which enable both high levels of abstraction and categorical thinking, which collapse in tandem with injury, and which the surviving parts of the organism, "all its powers," seek to reestablish and reconstruct.[53] Individual patients, having lost the original wholesome integration, attempt through these "powers" to produce new conscious integrations—in the process failing in different ways to overcome verbal, syntactical, nominal, or semantic losses or disorders. In his linguistic unfreedom the aphasic patient becomes pathologically individualized.

## Goldstein: Aphasia, Tonus, Organism

AS WITH HEAD, Goldstein's major work in aphasia began during the war; it culminated in the publication of a series of highly detailed cases studied with Gelb in the *Psychologische Analysen hirnpatholo-gischer Fälle*. Goldstein recorded extensive accounts of the range and consequences of injury, most famously in the case history of a twenty-four-year-old former miner named Johann Schneider (now mostly known by his last name), who on June 4, 1915, was struck in the occipital lobes by two shards of mine shrapnel—one wound above the left ear, the second in the "middle of the back of the head . . . penetrating into the exposed brain."[54] As Anne Harrington has noted, Schneider became a veritable Anna O for Gelb and Goldstein; indeed he would be treated as such by both Ernst Cassirer in *Philosophy of Symbolic Forms* and Maurice Merleau-Ponty in *The Structure of Behavior* and *Phenomenology of Perception*.[55] In a 140-page study Gelb and Goldstein argued that even though Schneider passed the traditional tests aiming to point out reduced capacities (tests ostensibly premised on atomistic symptomatology, reflex psychology, and localization), upon closer observation he was incapable of a natural performance of most of the tasks he had managed in tests. Gelb and Goldstein highlighted a series of failures, which led Schneider to attempt to reconstruct the lost performances differently. Reading, for example, required "a series of minute head- and hand-movements—he 'wrote' with his hand what his eyes saw. He did not move the entire hand as if across a page, but 'wrote' the letters one over the other, meanwhile 'tracing' them by head-movements. An especially interesting aspect of the case was the patient's own ignorance of using this method. . . . If prevented from moving his head or body, the patient could read nothing whatever. . . . If required to trace a letter the 'wrong' way, he was quite at a loss to say what letter it was."[56] Simply put, here was a patient who could carry out concrete tasks yet could not handle even basic uncontextualized mental operations except in an extremely belabored way (if at all), a patient who sought to compensate in his performances for the natural attitude he had lost, an attitude that Gelb and Goldstein called *abstract* or *categorial*.[57]

Rather than focus on the case studies, or on Goldstein's short work *Über Aphasie* (*On Aphasia*, 1926),[58] where he worked to dismantle existing symptomatology and localization frameworks, we will proceed directly to *The Organism* (1934), which furthered his research on sensorimotor consequences of neuronal damage and attempts to establish a complete theory of organismic functioning, grounded in neuropathology. We then return to reference certain among Goldstein's works that have been by and large overlooked, namely studies on tonus from the 1920s, which help explain his shift from the study of aphasia to an argument concerning the organism as a whole.

Central to Goldstein's argument in *The Organism* was less the localization of speech centers—which he had repeatedly criticized earlier and considered a lesser and widely accepted problem—than the refusal of *atomistic symptomatology*, the study of symptoms as separate and sufficient indicators of damage.[59] The book's structure is tied to Goldstein's sculpting of a broad alternative approach, which he could guide toward a rethinking of central medical notions, such as health, norms, disease, and cure.

In his theorization of diagnostic and therapeutic efforts already before *The Organism*, Goldstein charged that atomism went hand-in-hand with a thin comparative analysis that centered on *lesions* and individual failures rather than attending to the way the organism responds to general loss and attempts to compensate for the loss of functions. Atomism, and the localization theory underlying it, denied the tremendous complexity of pathological "performances" exhibited by the organism, in other words, the activities and responses to prompts that the organism carries out in everyday life, without breaking them down into component parts (reflexes). Nor did atomism, Goldstein insisted, offer a way such compensation might be turned to the patient's advantage.[60] The approach offered little by way of therapy for the brain-injured soldier, as it failed to capture the precise restriction of his abilities that resulted from irreparable damage to areas of brain tissue.

Atomism presumed that damages resulted in the same symptoms and disturbances, that individuals were affected in the same way, and that comparative analysis only replicated itself accordingly. By attending instead to each *individual* organism, Goldstein refused to see particular disturbances merely as facts, asking rather "what kind of a fact an observed phenomenon represents." In other words, at stake for the patient, the therapist, and the physiologist were not symptoms themselves nor the ways these were structured by a reflex-based integration, but what Goldstein referred to as performances. The organism's *original* "ordered behavior" was disrupted, and the organism found itself performing in a disordered fashion that fundamentally sought to restore itself to a new order:

> In an *ordered* situation, responses appear to be constant, correct, adequate to the organism to which they belong, adequate to the species and to the individuality of the organism, as well as to the respective circumstances. The individual himself experiences them with a feeling of smooth functioning, unconstraint, well-being, adjustment to the world, and satisfaction, i.e. the course of behavior has a definite order, a total pattern in which all involved organismic factors—the mental

and the somatic down to the physico-chemical processes—participate in a fashion appropriate to the performance in question. And that, in fact, is the criterion of a normal condition of the organism.

In this passage Goldstein deploys three of his core concepts, namely *wholeness, performance,* and *order.* First, it is the totality of the organism, *"all* involved organismic factors—the mental and the somatic down to the physico-chemical processes," and neither "merely" psychological nor "merely" bodily ones, that are involved. Second, Goldstein insists on the organism's "performances": it is "normal" performances that define health and order and that are impacted by disorder. Third, order is also fundamental: when it finds itself "dis-ordered," the organism naturally attempts to compensate, to achieve a new order. Furthermore certain kinds of *dis*order could even result in a *"catastrophic* reaction" wherein the organism finds itself failing through and through to handle any of the environment's requirements. In a passage where he intentionally conflates a disordered situation *within* with the presence of disorder *around* the patient, Goldstein writes, "The principal demands which 'disorder' makes upon [patients] are: choice of alternatives, change of attitude, and rapid transition from one behavior to another. But this is exactly what is difficult or impossible for them to do. If they are confronted with tasks which make this demand, then catastrophic reactions, catastrophic shocks, and anxiety inevitably ensue. To avoid this anxiety the patient clings tenaciously to the order which is adequate for him, but which appears abnormally primitive, rigid and compulsive to normal people."[61] Order and disorder, compensation, and anxiety are central points in the general biological and physiological effort to counter atomism and, at the neurological level, reflex theory itself.

In *The Organism,* Goldstein critiques Sherrington's *Integrative Action of the Nervous System* extensively, making an obvious effort to replace it as the neurophysiological standard.[62] This critique was central to Goldstein's argument: for Goldstein reflex theory posited atomistic responses, that is, responses that would be possible to subdivide to the point of being able to deal with their components as individual, simple reflexes; reflex theory further posited that responses to identical stimuli were themselves going to be *identical*, but reflexes, Goldstein argued, are themselves milieu-based and not of equal consequence in different contexts.[63] Thus, Goldstein charged, physiology became fundamentally reductive, with researchers testing local reflexes or reflex arcs rather than remembering that it was not these that hurt the patient but the complex performances that he had lost and sought to regain.[64] He did not see disease or disturbance as the sole origin of a disruption of behavior; rather *the* crucial determinant was the organism's adjustment, within its given milieu, to such a disturbance. Disease and disturbance forced the organism to an adaptive, conservative situation where its disordered performance tended toward compensation or "readjustment" in order to generate a new order, a new normality. Specifically with regard to central cortical injury—including the aforementioned cases that resulted in visual and linguistic disturbance, Goldstein offered a further point, namely that their effect was to destroy patients' capability for abstract thought and leave them able to respond only to immediate concrete circumstances. Compensation for disease and disturbance caused the organism to lose its well-ordered *abstract* or *categorial* capacity:

> Whenever the patient must transcend concrete (immediate)
> experience in order to act—whenever he must refer to things
> in an imaginary way—he fails. On the other hand, whenever
> the result can be achieved by manipulation of concrete and
> tangible material, he performs successfully. Each problem that
> forces him beyond the sphere of immediate reality to that of
> the "possible," or to the sphere of representation, ensures his
> failure. This manifests itself in all responses. . . . The patient
> acts, perceives, thinks, has the right impulses of will, feels
> like others, calculates, pays attention, retains, etc., as long as

he is provided with the opportunity to handle objects concretely and directly. He fails when this is impossible. . . . He
can manipulate numbers in a practical manner, but has no
concept of their value. . . . He is incapable of representation of
direction and localities in objective space, nor can he estimate
distances; but he can find his way about very well, and can
execute actions that are dependent upon perception of distance
and size. . . . The patient has lost the capacity to deal with *that
which is not real—with the possible.*

Categorial behavior, "an attribute of the human being," he writes
later in the book, involves "the ability of voluntary shifting, of
reasoning discursively, oriented on self-chosen frames of reference, of free decision for action, of isolating parts from a whole, of
disjoining given wholes, as well as of establishing connections."[65]
In opposition to the categorial attitude, which for him "embraces
more than merely the 'real' stimulus in its scope" and which is
characterized by a grasp of order, a wholeness and normality of
experience, the concrete attitude is "realistic" and "does not imply
conscious activity in the sense of reasoning, awareness, or a self-
account of one's doing."[66]

Crucially the loss of the abstract or categorical attitude in brain
injury was central to the elimination of the old order and to the
reduction of the patient to a disorderly situation where all he
could cope with was the immediate, concrete situation surrounding him. The cortically injured patient faces "systemic disintegration" rather than the mere loss of particular performances,
as those inevitably affect the organism's performance of wholeness. Faced with abstract demands, a patient would begin to sense
intense anxiety at his inability to fulfill them and would emphatically seek to return to some order. These "catastrophic" situations may involve a further reduction in the capacity to cope, and
the organism, operating without abstract or discursive thought,
performs merely with a *restitutio ad integrum* in mind—however
restricted that *integrum* might be.[67]

Note how the apparently slight difference of priorities and targets comes to play a significant role in the milieu Goldstein located in the patient as well as the healthy organism. Unlike Head, Goldstein had no difficulty arguing "from the pathological to the normal,"[68] and in fact he presented the extrapolation from pathological to normal organisms as the basis of his thought. Where Head focused on the new integrations actively carried out by the injured patient, distinguished those from "normal" language use, and left it largely to experimental neurology to retrieve the bases of "the normal" body, Goldstein proceeded directly to a theory of the structure of *the organism* from his pathological cases.

Goldstein further distinguishes himself from Head by attending at length to the motor capacities and posture of the organism, in particular the problem of tonic musculature. At every moment of our lives the tonus of our musculature is regulated, and without this constant adjustment we could neither sit nor stand nor walk. But these adjustments to our tonic form take place in a completely involuntary fashion, without our knowledge or cognition. In particular Goldstein showed the importance of head posture (and its modifications) in understanding brain lesions, and he would later be credited with the discovery that tonic neck reflexes "disclosed first the existence of an extended motor relationship between different parts of the body."[69]

In a number of studies written between 1924 and 1930 and recounted in *The Organism* as well as in a series of films he made, Goldstein described in detail a series of specific tests, which pertain to the body's tonic musculature and are aimed at understanding the tendency to compensate for injury so as to achieve some optimization, however difficult or tenuous.[70] Scholars have attended to the optical tests administered by Goldstein and Gelb, but it is in these studies of tonus that the point of "performances" and the extrapolation of neurological concerns toward organismic wholeness (in fact several major contributions of Goldstein's) become clear.[71] Concerned with accounting for the patient's experience of environment, Goldstein focused on more than "problem solver" experiments and rejected views that dispensed with the environment as mere "background" from which the organism remained independent. Thus for the psychologist Kurt Koffka, for example, "there is no behavioral environment for [tonus]."[72] But where Koffka retained a reflex paradigm, Goldstein saw tonic musculature and especially *changes* in tonus as a modification in the individual on the total neuromotor level, a modification that *incorporated* (the word is apt) the environment. These changes were fundamental to Goldstein's attempt to account for what disordered modifications meant for *this* individual given *that* demand.[73] Tonic change in one area of functioning would often suggest a systemic change in the entire organism's functioning, and whereas neurobiological vigilance in Head's case gave support to a reflex-like responsive protopathic handling of the environment, for Goldstein tonus marked a deeper ingraining of the environment into the very structure of a disordered or partially reordered organism. In *The Organism* he discussed compensation and optimization at length in relation to organismic responses to new contexts and to the "preferred behavior" in which they resulted:

> It seems that the adaptation to an irreparable defect takes essentially opposite directions. *Either* the organism adapts itself to the defect, . . . yields to it, . . . resigns itself to that somewhat defective but still passable performance which can still be realized, and resigns itself to certain changes of the milieu which correspond to the defective performances; *or* the organism faces the defect, readjusts itself in such a way that the defect, in its consequences, is kept in check. . . . In patients with one-sided cerebellar lesion, we often find a "tonus pull" towards the diseased side. All stimuli which are applied to this side are met with abnormal intensity, with abnormal "turning to the stimulus." This leads to deviation in walking, to a predisposition to falling, to past pointing, etc., all towards

the diseased side. . . . As long as the patient remains in this abnormal posture he feels relatively at ease, has less subjective disturbances of equilibrium, less vertigo, etc. . . . Deviations may disappear completely. . . . Disturbances immediately reappear as soon as the patient reassumes the old, normal position of the body. Apparently the abnormality of posture has become the prerequisite for better performances, has become the *new preferred situation.*[74]

The subjective/objective division matters: in "objectively" abnormal behavior, we see that the modified responses and deviations represent a profound discomfort at the supposedly normal, "objective" demands that have now been rendered meaningless. Instead a new normal has arisen, specific to the patient's restricted perspective, and these deviations indicate the reconstructed and indeed preferred—in many respects healthier—perspective. In the world, patients would often find ways to compensate for losses or changes in functions, sometimes without notice, and it was only in a crisis that arose from no longer being able to meet the demands of everyday life that a state of injury could be called pathological (fig. 5.4).

Tonic change as a *loss* of some capacity was at that point pathology, but in experimental work on tonic movement, Goldstein could demonstrate modifications in the capacity of the body to perform motor tasks *with or without* such loss in the world, giving some clue as to how the environment of capacities had been narrowed but also how it was being manipulated anew by the modified individual. In what he discusses above as "tonus pull," we find not only the apparent defect but also the sense in which the individual has compensated and reordered the world (fig. 5.5).

This bears emphasis: in his *Philosophy of Symbolic Forms*, Cassirer provides a summary of the clinical scene at the Frankfurt Neurological Institute, yet he retains only the pejorative sense of disorder and not the productive sense of the organism's compensation.[75] Goldstein, by contrast, emphasized the organism's fundamental tendency toward *preferred behavior*, which is "determined by the total attitude

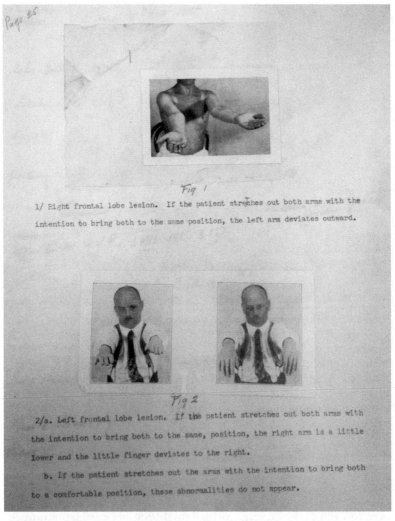

**Figure 5.4**  Tonus changes and abnormalities due to frontal lobe lesions. Columbia University, Rare Book and Manuscript Library, Kurt Goldstein Papers, Box 6, "Frontal Lobal Lesions" folder, p. 25.
Source: Kurt Goldstein Papers, Rare Book & Manuscript Library, Columbia University in the City of New York.

of the performing person." In cases of injury a new "preferred behavior" emerges to basically generate the greatest comfort.[76] In the case of patients reported as cases or appearing in his research films, notably *Tonus* (1930), we can see this preferred behavior expressed nega-

**Figure 5.5** As the patient closes his eyes, he loses his spatial sense, so that any turn of the head produces an involuntary movement of the arm (in this case the right arm). From *Tonus* (1930). Columbia Rare Book and Manuscript Library, Kurt Goldstein Papers, Box 18. *Tonus* is one of five films surviving in Goldstein's archive, some of which he brought with him when he emigrated, for the purposes of illustrating lectures and convincing American institutions of the particularity and originality of his work. See Robert A. Lambert to Collector of Customs, October 15, 1934, Rockefeller Archive Center, RF/1.1/200A/78/939.
Source: Kurt Goldstein Papers, Rare Book & Manuscript Library, Columbia University in the City of New York.

tively through groups of symptoms exhibited when "normal" actions are required of the patient: "unilateral disequilibrium, "spontaneous inclination" (leaning and swaying) of the body, the passive turn of the head, further aggravated body and arm deviation, and so on. Taken as a whole these symptoms show a body struggling against the demands of seemingly minimal tasks; they also indicate a "preferred" posture and stance, a sense of "subjective" normalcy at odds with our sense of "normal" behavior during such tasks (fig. 5.6).

In addition to grasping and pointing tests, which allowed Goldstein to observe movement and neuromuscular response, he

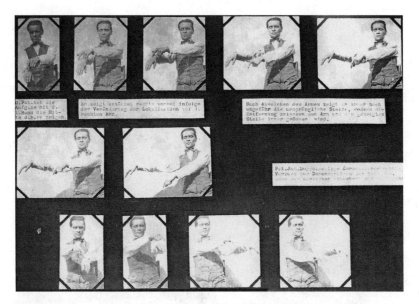

**Figure 5.6**  Presentation slide involving the same patient attempting a pointing test. Columbia Rare Book and Manuscript Library, Kurt Goldstein Papers, Box 18, Slides.
Source: Kurt Goldstein Papers, Rare Book & Manuscript Library, Columbia University in the City of New York.

incorporated tests for assessing equilibrium and sensory response with the aid of instruments as simple as a tuning fork, a violin, or a set of keys.[77] In *Tonus* he uses air puffs through a tube to create pressure—a common test for lesions of the inner ear used to assess percussion injuries and their effects, especially on equilibrium and dizziness (fig. 5.7).[78]

There are several motivations for using this test, which point to the specificity of Goldstein's interests and the physiological and psychological relevance of the experiments. For example, the failure to react to ear stimulation (due to lesions that fail to transmit sensation) was most associated with cerebral tract lesions. But this test, like others that Goldstein carried out—including spontaneous pointing or making heels and toes touch while keeping one's eyes closed—were tests used to assess the symptoms of disequilibrium and vertigo in a broader range of pathologies than brain injury

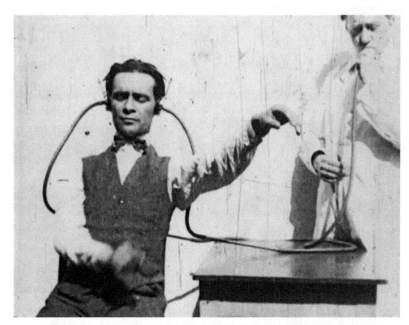

**Figure 5.7**  Use of ear puffs in *Tonus* (1930). The patient is instructed to
hold his arms steady and then, with closed eyes, to repeatedly touch the
top of his left hand with his right. With pressure applied in the inner ear,
the left arm begins to veer leftward, and the right one fails over and over
to touch it. Goldstein in the background.
Source: Kurt Goldstein Papers, Rare Book & Manuscript Library, Columbia
University in the City of New York.

alone.[79] To Goldstein they concerned the total comportment of the
organism: from the study of brain injury he had moved to the neuromuscular transformation of the organism in its environment.

In other words, it was not despite but because of these similarities between inner-ear damage, vertigo, disequilibrium, and brain
lesions that Goldstein sought to test a range of symptoms. He writes,
"We soon found that one reason for failure in treatment was that
we overlooked the fact that similar-appearing symptoms can be of
essentially different origins, and that only by knowing that latter
can one avoid inadequate treatment and achieve better results."[80]
To examine and understand the patient's loss of order and abstract
attitude, it became imperative to test symptoms in groups and seek

the cause of performative failures. Together the results of these tests produced *facts* not about the whole of patients under examination but about the whole of the single patient, his capacities, and the difference between his "preferred" behavior and the "normal" one.

IT IS NOT the point of this essay to split hairs between Head and Goldstein. We wish instead to focus on the way that divergences in approach and argument had profound consequences for the otherwise proximate conceptions of the self (healthy or injured), of disease and rehabilitation, of society, language, and milieu, and of the mind-body relation that the two authors proposed. *Nevertheless* the logic binding integration to disequilibrium remained a dominant imperative in the work of each of them: their handling of order, integration, catastrophe, and related concepts displays the centrality of this logic to their thought. Not only do the two authors raise the question of how an organism exists as a single whole and tends again and again toward wholeness during and after injury, but they establish this collapse and the question of wholeness differently. The different levels of Head's integration—reflexive, neurological, linguistic—and his focus on identifying and restoring linguistic difficulties differ from Goldstein's holistic and psychological sense of an organism aiming at self-actualization but losing its abstract capacity and its unitary and normative capacities. More broadly the two authors stood far apart as regards man's engagement with language, as regards monism and dualism and thus the status of the mind, and as regards the facts and concepts of disease and recuperation. In the remaining pages we consider the implications of the two arguments by sculpting each in the other's relief.

Head's conception of integration occurs within a two-tiered system. The first tier is the sense of a basic neural and corporeal integration prompted by the Sherringtonian account of nervous integration, reminiscent of protopathic sensibility, and elaborated in the argument on *vigilance*. We might call this Head's conception of the *integratedness* of the organism; its integration is a given, something that takes place deep beneath any consciousness and appears as a given even in cases of decorticated animals. The action and reaction of impulses, including reflexes, and more broadly "vigilance" (once again a concept he abstracts out of neurology and not from clinical work) are essential for any "psychoneural" integrated movement of the organism. A basic integration at the physiological and unconscious levels is something that exists and is *limited* only by the loss of particular lesions; for anything whatsoever to happen in the organism, such vigilance and its restoration remain essential, sensed but unaccounted for.

For the second tier, *on top of and coefficiently* with the first, Head makes much, in both his theoretical and clinical descriptions, of *active* integrative forces, which he foregrounds as occurring in the aftermath of injury. As quoted earlier, he writes that injury is followed by "a fresh integration carried out by all available portions of the central nervous system. It is a total reaction to the new situation."[81] To compare: Goldstein's *Organism* by and large describes the patient in terms that indicate him or her to be lacking freedom, restricted, tied to a "shrunken milieu," struggling between disorder and a new restricted and highly precarious order, constantly threatened with catastrophe; for Head the injured patient usually recovers to a considerable, even remarkable degree.[82] The difference from Goldstein is one of tone, but still very significant, so that Head provides a strongly active sense of consciousness *integrating* when he writes that, "faced with a new situation, the organism puts out all its powers, conscious, subconscious and purely physiological, in order to produce an adequate response directed towards its welfare as a whole," or when he sees consciousness as "a form of integrative vital reaction which enables the organism to adapt itself more perfectly to certain situations, conditioned by its internal state and the impressions produced upon it by external forces."[83]

Goldstein's theorization involves an integration described along two *non*tiered, superimposed schemata. In the abstract/concrete schema brain injury invariably amounts to the destruction of the abstract or categorial capacity of the organism and results in a concretization of the organism, an inability to plan, to extract oneself from immediate circumstances.[84] For Goldstein the brain-injured patient *loses* his abstract or categorial attitude and does not regain it. Goldstein minces no words, and though Head says something very similar on occasion, indeed although a parallel can be drawn to Head's protopathic and epicritic sensation, the centrality and priority in Goldstein of this destruction of the abstract capacity can be contrasted with the occasional, ephemeral place it holds in Head.[85] Thus Head swings as if on a pendulum between an argument that there is a fundamental diminution and that this diminution does not destroy the whole and can perhaps be regained,[86] but Goldstein simply sees the loss of abstract attitude as devastating, as imprisoning the patient into the concrete immediate situations in which he finds himself. Language and performance are apiece with abstraction here, and true integration resides only in the whole, healthy, normal organism, whereas in Head language can be studied in the restricted and disrupted wholes that patients develop when recuperating. For this argument Goldstein's focus on performances—on the natural execution and continuous, uninterrupted, untroubled performance of particular tasks, often not recognized as tasks—is essential. In seeing little of interest in Sherrington's integrative action or in Head's concept of vigilance as the basic neurobiological regulator, and in focusing instead on the relation of the subject to his milieu, Goldstein conceives of life and individuality in terms of a holism that is constantly threatened with disintegration.

Goldstein also uses the structure of order/disorder/catastrophe in order to establish a sense of what sort of rehabilitation the physician may hope for in his patient. Here Goldstein's argument on norms is crucial: the healthy organism is ordered *to the point of never recognizing or sensing* the existence of this order; he is normative, free, so to speak, from order, fundamentally autonomous and *driven to* "self-actualization."[87] (By contrast, freedom in Head is freedom to use language without constraint.) The disorder that occurs with injury or disease involves at once the patient's and physician's recognition that an order did exist and that it has been lost; the management of disorder may perhaps lead to new orders, but it also involves the patient's fall to the purview of norms and a depreciation vis-à-vis these norms: "A life that affirms itself against the milieu is a life already threatened."[88] Autonomy is dramatically lessened and self-actualization is hampered; it can be pursued now only via the scene that the physician makes possible. Catastrophe, "systemic disintegration," involves not only immediate collapse but also, once recuperation begins, a further restriction of the milieu; hence Goldstein's claim that the organism in a catastrophic situation performs merely with a *restitutio ad integrum* in mind, however radically restricted that *integrum* might be.[89]

Goldstein's two schemas imply one another without being coextensive: "order" does not concern merely circumstances of brain damage, and a certain order can be regained where the abstract attitude cannot. It is on the basis of these two schemas that Goldstein can focus on articulating "self-actualization" as something that occurs in the injured organism as well as the healthy, as a value and ideal for the healthy individual as well as a purpose to restore to the injured or sick organism. Goldstein's epistemology and philosophical anthropology in *The Organism* are fundamentally based on the claim that the individual strives for wholeness. It is not given whole; when healthy it does not even have to experience wholeness, and it can always be broken up. It is constantly threatened. But life here is a relation to the production of oneself as nonthreatened, autonomous, normative, an individual in the sense of in-dividual, in-divisible. Life, even animal life, maintains itself by instituting its own norms, by seeking to maintain its individuality: "[Animal behavior] points . . . to an individual organization, on which basis alone it becomes intelligible as the expression of the tendency to actualize itself according to the circumstances."[90] Decomposition, or systemic disintegration, by contrast, was the gravest danger for this being.[91]

The stage is thus set very differently: the course of disease—and especially of consciousness in disease—like the place and centrality of an active integration in the diseased or disordered subject, are set in different terms. Where Head relies on vigilance for the original, healthy, normal integration, Goldstein turns to the "abstract attitude." Where Goldstein sees the abstract attitude (and the integration associated with it at the level of consciousness) as disintegrating and impossible to restore, Head sees no such attitude, only a neural, physiological integration beneath consciousness. Where Head tracks the patient's gradual improvement into language, treating it as a fresh integration, Goldstein instead sees it as highly unlikely, valorizes it as self-actualization, and identifies no grand restitution of the whole. If the degree zero of integration resided for Goldstein in the abstract attitude, for Head it did so in vigilance: integration and disequilibrium are structured around these concepts.

Head's turn to the neogrammarian theory of language further accentuates the very different kind of subject from the one that emerges from Goldstein's extrapolation of order and self-actualization out of tonus. As we have seen, Head posits that integration occurs at several levels—local and regional neurological levels, as well as both conscious and merely physiological levels—and these multiple integrations bear profoundly on language and its use: "The fact that speech can be disordered by injury to the brain is one of the most wonderful means placed in our hands for investigating the relation of mind and body."[92] Vigilance constitutes the hinge between them: any substantial reduction in its "neural potency" amounts to a destruction of the joint between the realm of consciousness, speech (interior and exterior) and language, and its neural underpinning. At this point "the mind/body problem does not exist so long as we are examining the consequences of functional disintegration"; in other words, *the organism becomes one in its attempt to create a "fresh integration,"* in its use of "all available portions of the central nervous system" to achieve it.[93] Language, and symbolic expression more generally, was only *partly* unavailable to these patients; they remained broken and lacked this "normal" access to the outside world. But even as reduced wholes, they experienced a substantial recuperation that went toward actively building a (new) organism and self.

This raises the difficult problem of monism and dualism—so central to psychology, physiology, and philosophy—which Head and Goldstein negotiated in quite different ways. In his more theoretical writings Goldstein moved toward a theory of self-actualization and holism—a half-hearted monism whose pursuit in the individual is *undermined* by brain injury, disorder, and the complete loss of the abstract attitude.[94] He saw monism in self-actualization, in the fully healthy individual seeking out, without interruption by the environment, the life he or she sought. Having escaped death only to lose their self, Goldstein's patients remained shattered into pieces, lost their chance at wholeness, became threatened with complete disintegration, and attempted out of their concrete experience to develop a reduced order.

But as we have seen, Head proposed instead that at stake were the particular ways in which the linguistic dimension of the self collapsed—and with it all direct access to language and society—so that the self would collapse into two. Losing the normal form that allowed its singularity, the coexistence and coextension of integrations, this broken self remained vigilant beneath consciousness, and it also began its pursuit of an internal reconstruction, sought a new wholeness without normal selfhood being quite available. Head's patients broke and then reconstructed a certain wholeness that crisscrossed language, a wholeness that retained them in perpetual tension with their environment and the demands of the linguistic order, that could not be easily or perhaps even adequately understood from the physician's linguistic perspective, and that could only be treated to aid toward a new and reduced harmony with the world. In other words, their pathological self-reconstruction was profoundly monistic, the self becoming one out of the fragments as it sought the new integration. Goldstein's patients were accorded no such trend-toward-unity by their neurologist.

This means, finally, that the two thinkers developed very different conceptions of individuality. For Goldstein only the healthy and free individual was truly individual—and he was individual precisely because he was free and healthy. For Head, by contrast, individuality resided in the pathological reintegration. It was precisely because each patient was a case unlike any other, with only barely any capacity for categorization into language and society, and at the same time because he was carrying out reintegrating processes while remaining unfree, imprisoned by his injury, that he could become truly, differently, pathologically individual.

# Endnotes

1   This essay is based on research at the Kurt Goldstein Papers, Rare Book and Manuscript Library, Columbia University; the Henry Head Papers at the Wellcome Institute Library; the Archives de Georges Canguilhem at the Centre d'Archives de Philosophie, d'Histoire et d'Edition des Sciences of the École Normale Supérieure; the Max Planck Institute for the History of Science library; and the Humanities Library of the Freie Universität Berlin. Special thanks are due to Emily Holmes and Jennifer Lee at Columbia.

2   "Neurasthenia and Shell Shock," *Lancet*, March 18, 1916, 627.

3   Henry Head, *Aphasia and Kindred Disorders of Speech*, 2 vols. (1926; London: Hafner, 1954), 1: 140.

4   Kurt Goldstein, *After-effects of Brain Injuries in War* (New York: Grune & Stratton, 1942), 13. Goldstein explains the need for long-term, close observation in *The Organism* (New York: Zone Books, 1995), 40–41.

5   See, e.g., Karl S. Lashley, "Basic Neural Mechanisms in Behavior," *Psychological Review* 37 (1930): 1–24.

6   Head, *Aphasia and Kindred Disorders of Speech*, 1: 63.

7   Charles S. Sherrington, *The Integrative Action of the Nervous System*, 2nd edition (1906; New Haven, CT: Yale University Press, 1947).

8   Mitchell Ash, *Gestalt Psychology in German Culture, 1890–1967: Holism and the Quest for Objectivity* (Cambridge: Cambridge University Press, 1998). See Head's critique of this Spencerian psychology in Hughlings Jackson: Henry Head, "Hughlings Jackson on Aphasia and Kindred Affections of Speech," *Brain* 38 (July 1915): 1–2.

9   Sherrington, *Integrative Action of the Nervous System*, 7.

10  Ibid., 313, 352.

11  See the detailed biography by L. Stephen Jacyna, *Medicine and Modernism: A Biography of Henry Head* (London: Pickering & Chatto, 2008).

12  Kurt Goldstein, *Die Behandlung, Fürsorge und Begutachtung der Hirnverletzten* (Leipzig: Vogel, 1919); Kurt Goldstein, *Über Aphasie* (Leipzig: Orell Füssli, 1927), originally published in *Schweizer Archiv für Neurologie und Psychiatrie* 19, nos. 1–2 (1926).

13  Head, *Aphasia and Kindred Disorders of Speech*, 1: 141, also 1: 129–33. See Adhémar Gelb and Kurt Goldstein, "Zur Psychologie des optischen Wahrnehmungs- und Erkennungsvorganges," in *Psychologische Analysen hirnpathologischer Fälle* (Leipzig: J. A. Barth, 1920); Adhémar Gelb and *Goldstein*, "Über *Farbenamnesie* nebst Bemerkungen über das Wesen der amnestiaschen Aphasie überhaupt und die Beziehung zwischen Sprachen und dem Verhalten zum Umwelt," *Psychologische Forschung* 6 (1924): 127–86.

14  Kurt Goldstein to Henry Head, July 23, 1926, Wellcome Institute Library, PP/Hea/D8.

15  See Cassirer's request for a reading list: Cassirer to Goldstein, January 5, 1925, in Ernst Cassirer, "Two Letters to Kurt Goldstein," *Science in Context* 12, no. 4 (1999): 663. See also Cassirer's references to Head in *Philosophy of Symbolic Forms*, vol. 3: *The Phenomenology of Knowledge*, trans. Ralph Manheim (New Haven, CT: Yale University Press, 1957), chapter 6. Nevertheless either Goldstein did not possess significant correspondence with Head, or he decided to leave it behind when he was forced to emigrate from Germany in 1933. (Very few letters from before 1933 survive in Goldstein's archive, the ones from Cassirer being one notable set.)

16  Kurt Goldstein, untitled, undated lecture from the 1950s, Columbia University, Rare Book and Manuscript Library, Kurt Goldstein Papers, Box 3, folder "Aphasia," p. 5.

17  Jacyna, *Medicine and Modernism*.

18  Head, *Aphasia and Kindred Disorders of Speech*, 1: viii. Head notes that he first encountered aphasia at Addenbrooke's Hospital at Cambridge in 1886 (a decade before he moved to London Hospital). Goldstein's earliest essays on aphasia are a case study dating to 1906 and a commentary on François Moutier's commentary on Broca's area. See Kurt Goldstein, "Zur Frage der amnestischen Aphasie und ihrer Abgrenzung gegenüber der transcorticalen und glossopsychischen Aphasie," *Archiv für Psychiatrie und Nervenkrankheiten* 41, no. 3 (1906): 911–50; Kurt Goldstein, "Einige Bemerkungen über Aphasie im Anschluss an Moutier's 'L'aphasie de Broca,'" *Archiv für Psychiatrie und Nervenkrankheiten* 45, no. 1 (December 1908): 408–40.

19  Institut für die Erforschung der Folgeerscheinungen von Hirnverletzungen. See Frank W. Stahnisch and Thomas Hoffmann, "Kurt Goldstein and the Neurology of Movement During the Interwar Years," in *Was bewegt uns? Menschen im Spannungsfeld zwischen Mobilität und Beschleunigung*, ed. Christian Hoffstadt et al. (Bochum: Projekt Verlag, 2010), 288.

20  On Rivers's importance to the establishment of psychology in Britain, see, among others, John Forrester, "1919: Psychology and Psychoanalysis, Cambridge and London," *Psychoanalysis and History* 10, no. 1 (2008): 37–94. For the reports on the Torres Straits Expedition, see Alfred Cort Haddon, W. H. R. Rivers, C. G. Seligman, Charles S. Myers, William McDougall, Sidney Herbert Ray, and Anthony Wilkin, *Reports of the Cambridge Anthropological Expedition to Torres Straits*, 6 vols. (Cambridge: Cambridge University Press, 1901–35).

21  Note Canguilhem's rather different interpretation of the relationship of the two: "The conceptions of Hughlings Jackson, Head and

Sherrington, paving the way for more recent theories such as those of Goldstein, oriented research in directions where facts took on a synthetic qualitative value, at first ignored." Georges Canguilhem, *The Normal and the Pathological* (New York: Zone Books, 1991), 86.

22 There are significant exceptions to this: both Head and Goldstein praise Constantin von Monakow and Arnold Pick in addition to Hughlings Jackson, though Head's critical argument was much more oriented toward criticizing them for what they did not manage.

23 Head republished Hughlings Jackson's principal essays on aphasia in 1915, in *Brain* 38, nos. 1–2 (1915). Note the regard he expresses for Hughlings Jackson in his (and G. Holmes's) 1911 Croonian Lecture: "Others have been our teachers and the masters for whom we have gladly worked. To them our debt can never be repaid, for much of the identity in outlook comes from personal intercourse and cannot be fixed by citation. First amongst this group must always stand the name of Dr. Hughlings Jackson." As published in Henry Head and Gordon Holmes, "Sensory Disturbances from Cerebral Lesions," in Henry Head, *Studies in Neurology* (London: Hodder and Stoughton, 1920), 2: 533. On Hughlings Jackson, see Nima Bassiri, this volume.

24 Head, "Hughlings Jackson on Aphasia and Kindred Affections of Speech," 1.

25 Head, *Aphasia and Kindred Disorders of Speech*, 1: 133.

26 Ibid., 1: x.

27 Ibid., 1: 134, ix.

28 Given his polemical predilection, Head is actually remarkably kind to Gelb and Goldstein; he does not target their use of terms and notions (e.g., word-blindness) whose use in other figures he objects to strenuously (see the critique of Bastian and Kussmaul over word-blindness in ibid., 1: 117).

29 Ibid., 1: 197, 133.

30 Head presents the clinical material underlying the argument at far greater length in the second volume, warning that even these case details are "drastically reduced versions of voluminous clinical records" (ibid., 2: vii). The first seventeen cases, clearly placed first in order to prioritize wartime injury, are all of men who came under Head's care during the war, suffering from brain injury due to gunshots or other fragments; the remaining nine cases, of seven men and two women, are almost all patients age fifty-five to sixty-five and date with only one exception to the postwar. (The one exception is case 21, who was brought to Head in 1910; the patient of case 25 also visited Head before the war, but he attended to her, he says, with much greater care in 1920, and his notes date from that period.)

31 Head, *Aphasia and Kindred Disorders of Speech*, quoted in Columbia University Rare Book and Manuscript Library, Kurt Goldstein Papers, Box 3, "Aphasia" folder, page numbered "5."

32 Head, *Aphasia and Kindred Disorders of Speech*, 1: 138.

33 Ibid., 1: 218. See the strong critique of the approaches he described as prevalent at the end of the war (1: 202).

34 Ibid., 1: 210–18.

35 Marcel Mauss, "Rapports réels et pratiques de la psychologie et de la sociologie," *Journal de Psychologie Normale et Pathologique* 21 (1924): 892–922, republished in *Sociologie et anthropologie* (Paris: PUF, 1950), 281–310.

36 Head, *Aphasia and Kindred Disorders of Speech*, 1: 269, 323, 203, 220.

37 Ibid., 1: 289.

38 Ibid., 1: 289, almost identically phrased in 2: x.

39 Ibid., 1: 269, 270.

40 Ibid., 1: 544, 535.

41 Ibid., 1: 166.

42 Ibid., 1: 220, first discussed in Henry Head, "Aphasia and Kindred Disorders of Speech," *Brain* 43, no. 2 (1920): 119–57; Henry Head, "Disorders of Symbolic Thinking and Expression," *British Journal of Psychology* 11, no. 2 (1921): 179–93.

43 Head, *Aphasia and Kindred Disorders of Speech*, 1: 229, 239, 241, 257; see also 261, 268 and 1: 456.

44 H. Charlton Bastian, *A Treatise on Aphasia and Other Speech Defects* (London: Lewis, 1898).

45 See the discussion of "the grammarians" in Head, *Aphasia and Kindred Disorders of Speech*, 1: 120–21.

46 Ibid., 1: 122.

47 Ibid., 1: 140–41. Cf. L. S. Jacyna, *Lost Words: Narratives of Language and the Brain 1825–1926* (Princeton, NJ: Princeton University Press, 2000), 152.

48 See Henry Head and W. H. R. Rivers, "A Human Experiment in Nerve Division," *Brain* 31 (1908): 323–450, republished in Head, *Studies in Neurology*, 1: 225–28.

49 Henry Head, "The Conception of Nervous and Mental Energy II: Vigilance: A Physiological State of the Nervous System," *British Journal of Psychology* 14, no. 2 (1923): 133; Head, *Aphasia and Kindred Disorders of Speech*, 1: 486.

50 Head, "Conception of Nervous and Mental Energy II," 134–35; Head, *Aphasia and Kindred Disorders of Speech*, 1: 487.

51  Head, *Aphasia and Kindred Disorders of Speech*, 1: 539.

52  Head, "Conception of Nervous and Mental Energy II," 137; Head, *Aphasia and Kindred Disorders of Speech*, 1: 490.

53  Head, *Aphasia and Kindred Disorders of Speech*, 1: 535, 533, 538.

54  Gelb and Goldstein, "Zur Psychologie des optischen Wahrnehmungs- und Erkennungsvorganges," 9.

55  Anne Harrington, *Reenchanted Science: Holism in German Culture* (Princeton, NJ: Princeton University Press, 1999); Cassirer, *Philosophy of Symbolic Forms*, vol. 3, chapter 6; Maurice Merleau-Ponty, *Phenomenology of Perception* (London: Routledge, 1962).

56  Gelb and Goldstein, "Zur Psychologie des optischen Wahrnehmungs- und Erkennungsvorganges," 18, translated as "Analysis of a Case of Figural Blindness," in *A Source Book of Gestalt Psychology*, ed. W. D. Ellis (London: Routledge, Kegan, 1938), 318–19.

57  Kurt Goldstein, *The Organism* (1939; New York: Zone Books, 1995), 39, 43–44.

58  Kurt Goldstein, *Über Aphasie*, originally based on a lecture given in Bern on February 27, 1926. As Goldstein wrote to Head about *Aphasia and Kindred Disorders of Speech* only in July of that year (according to Jacyna, *Medicine and Modernism*, 143, it had been published by March), it is unclear whether the latter work would have played a role in Goldstein's critiques of atomistic symptomatology and localization, which had long been important themes to Goldstein.

59  Goldstein, *The Organism*, 203–7, 34–35. See also Kurt Goldstein, *Selected Papers/Ausgewählte Schriften*, ed. Aron Gurwitsch et al. (The Hague: Nijhoff, 1971), 155–56.

60  Goldstein, *The Organism*, 41–42.

61  Ibid., 34, 36, 28, 42, 46, 48, 54.

62  Ibid., chapter 2, especially 70, 74, 76, 78.

63  Ibid., 69. Head says something quite similar, *but not in critique of Sherrington*. See Head, "The Conception of Nervous and Mental Energy II," 139; Head, *Aphasia and Kindred Disorders of Speech*.

64  For the critique of reductionist physiology, see Goldstein, *The Organism*, 108–9, 261.

65  Ibid., 336, 43–44, 301.

66  Kurt Goldstein and Martin Scheerer, *Abstract and Concrete Behavior*, *Psychological Monographs* 53, no. 2 (1941): 2, 3.

67  Goldstein, *The Organism*, 45, 54–55. For the implications of this position, see also Georges Canguilhem, *Knowledge of Life* (New York: Fordham University Press, 2008), 132.

68  Goldstein, *The Organism*, 29. On this matter, see Canguilhem, *The Normal and the Pathological*, 184.

69  L. Halpern, "Studies on the Inductive Influence of Head Posture," in Marianne Simmel, *The Reach of the Mind* (New York: Springer, 1968), 75. See also Raoul Mourgue, "La Concéption de la neurologie dans l'oeuvre de Kurt Goldstein," *L'encéphale* 32, no. 1 (1937): 38–39; Goldstein, *The Organism*, 80–81, and the case Goldstein discusses on 177–78.

70  See Kurt Goldstein, "Über den Einfluss der Motorik auf die Psyche," *Klinische Wochenschrift* 3 (1924): 1255–60; Kurt Goldstein, "Über induzierte Tonus-Veränderungen beim Menschen. II. Über induzierte Tonus-Veränderungen beim Kranken," *Zeitschrift für die gesamte Neurologie und Psychiatrie* 89 (1924): 383–428; Kurt Goldstein and Walther Riese, "Über induzierte Veränderungen des Tonus. V. Kritisches und Experimentelles zur Auffassung des Vorbeizeigens," *Monatsschrift für Ohrenheilkunde* 58 (1924): 931–40; Kurt Goldstein, "Zum Problem der Tendenz zum ausgezeichneten Verhalten," *Deutsche Zeitschrift für Nervenheilkunde* 109 (1929): 1–61; Kurt Goldstein, "Über Zeigen und Greifen," in *Selected Papers/Ausgewählte Schriften*.

71  Goldstein's commitment to understanding systems of organismic functioning also distinguished his "holistic" method from others working in Gestalt psychology and focusing on "figure-ground" relationships so to develop a theory of perception and cognition. Wolfgang Köhler, *The Place of Value in a World of Facts* (1938; New York: New American Library, 1966), originally presented as the William James lectures at Harvard University, 1934; K. Koffka, *Principles of Gestalt Psychology* (New York: Harcourt, Brace & World, 1935).

72  Koffka, *Principles of Gestalt Psychology*, 50–52.

73  Kurt Goldstein, "Notes on the Development of My Concepts" (1959), in *Selected Papers/Ausgewählte Schriften*, 5–14

74  Goldstein, *The Organism*, 334–35.

75  Cassirer, *The Philosophy of Symbolic Forms*, 3: 243 (citing Gelb and Goldstein, *Psychologische Analysen*, 206ff., 226ff.), 217.

76  Goldstein, *The Organism*, 270, 266.

77  This is especially the case in another surviving film, *Diegelman*; see our *Experimente im Individuum: Kurt Goldstein und die Frage des Organismus*, (Berlin: August Verlag, 2014), 79–83.

78  Isaac H. Jones and Lewis Fisher, *Equilibrium and Vertigo* (Philadelphia: J. B. Lippincott, 1918), 221, 218.

79  Ibid., 227, 225.

80  Kurt Goldstein, "Das Wesen der amnestischen Aphasie," *Schweizer Archiv für Neurologie und Psychiatrie* 15 (1924): 163–75.

81  Head, *Aphasia and Kindred Disorders of Speech*, 1: 544.

82  Goldstein, *The Organism*, 56, 339. See also Canguilhem, *The Normal and the Pathological*, 185, 186; Head, *Aphasia and Kindred Disorders of Speech*, 1: 220–21.

83  Head, *Aphasia and Kindred Disorders of Speech*, 1: 535, 536.

84  See also Uta Noppeney, *Abstrakte Haltung: Kurt Goldstein in Spannungsfeld von Neurologie, Psychologie, und Philosophie* (Wurzburg: Konigshausen & Neumann, 2000).

85  Head, *Aphasia and Kindred Disorders of Speech*, 1: 538–39. As mentioned earlier, Head is explicitly dependent on Gelb and Goldstein's early publications for his infrequent uses of the term *abstract* in particular; for the derivation from Goldstein, see 1: 133, where Head accords Gelb and Goldstein the terms *abstract* and *concrete*.

86  Ibid., 1: 538–39, 535.

87  Goldstein, *The Organism*, 163.

88  Canguilhem, *Knowledge of Life*, 113.

89  Goldstein, *The Organism*, 45, 54–55.

90  Ibid., 333, 355.

91  For further references to "disintegration," see ibid., 33, 42, 44, 45, 115, 208, 341, 365, 370.

92  Head, "Disorders of Symbolic Thinking and Expression," 180. Head too considers, in a manner largely derivative of Sherrington, the tonus and posture problem, but he does so in an unsystematic fashion and certainly not with an aim to use it as central to the whole organism.

93  Head, *Aphasia and Kindred Disorders of Speech*, 1: 544.

94  For greater detail on the problem of monism and dualism in Goldstein, see our *Experimente im Individuum* 134–36.

*Hannah Proctor and Laura Salisbury*

# 6 The History of a Brain Wound: Alexander Luria and the Dialectics of Soviet Plasticity

*The "higher psychological functions" ought to have their own origin, but this origin should not be sought in the depths of the soul or in the hidden properties of nervous tissue; it should be sought outside the individual person, in objectively existing social history.*
—Alexander Luria, "L. S. Vygotsky and the Problem of Localization of Functions," *Neuropsychologia* 3, no. 4 (1965)

ON MARCH 2, 1943, a twenty-three-year-old Russian experienced a revolution every bit as transformative as the one that just preceded his birth. The revolution of 1917 had transformed the social world and had led to the creation of the Union of Soviet Socialist Republics in 1922. Coming to consciousness in the same historical conditions as the USSR itself, this young Russian was to remain tied both to its hopes and its disasters, for in the Battle of Smolensk, as a solider in the Red Army, he received a devastating bullet wound that penetrated the left parieto-occipital area of his cranium and created, in an instant, a revolution of the self. The injury and the scar tissue that altered the configuration of the lateral ventricles as the body attempted to heal itself left him with profound deprivations: his vision was deeply impaired; sometimes he would "lose" his sense of the right side of his body, while at others he thought that "his head had become inordinately large, his torso extremely small and his legs displaced";[1] he was unable to orient himself in space; his memory was damaged; and his ability to speak, read, and write was devastated. By his own and his doctor's account, he was a man whose world had been irreparably shattered.

The young man was Sublieutenant L. Zasetsky, the subject of Russian neuropsychologist Alexander Luria's famous 1971 "neurological novel" *The Man with a Shattered World: A History of a Brain Wound*. The book was written by Luria following twenty-five years of clinical observation and treatment, and for the philosopher Catherine Malabou, its account of Zasetsky's revolutionized subjectivity is deeply revealing. Malabou cites Zasetsky's sense of himself as "a kind of newborn creature" to support her conception of the explosive, destructive neural plasticity that can enter the self from without (as it did in this case, in the form of a bullet) but that is also immanent within the functioning of a brain ever vulnerable to strokes, tumors, and neurodegenerative disorders.[2] Malabou finds an unexpected political force in this destructive plasticity that she first articulates in *What Should We Do with Our Brain?* (2008). There she demonstrates that the concept of neural plasticity—the changes in neural pathways and synaptic activity that occur in relation to experience and that demonstrate that the brain is not a physiologically static organ—is all too frequently represented as a "brain fact" perfectly molded to the neoliberal formations of late capitalism. Donald Hebb's 1949 description of synaptic plasticity states that "any two cells that are repeatedly active at the same time will become 'associated,' so that activity in one facilitates activity in the other."[3] Plasticity, by this account, implies that brain function cannot be reduced simply to itself; instead it is formed in relation to experiences and elements extrinsic to it. As Nikolas Rose and Joelle M. Abi-Rached have recently put it, neuroplasticity suggests "an openness" of the "molecular processes of the brain to biography, sociality, and culture, and hence perhaps even to history and politics." But this openness of the very structures of neurobiological matter to the unpredictability of experience has nevertheless tended to be understood according to specific shapes of selfhood that emphasize liberal individuals' accountability in relation to their (and their offspring's) neurobiological development. Conceptions of the plastic brain have become molded and fixed according to a particular model of human nature in which our very neurobiology becomes "a site of choice, prudence, and responsibility for each individual."[4] As Jan Slaby and Suparna Choudhury note in their polemical "Proposal for a Critical Neuroscience," "scientific enquiry tends to mobilize specific values and often works in the service of interests that can easily shape construals of nature and naturalness";[5] therefore perhaps it is no surprise that neuroplasticity has found itself pressed into being at one with the formation of liberally responsible individuals and the bounded, directed flexibility demanded by an ever-changing flow of market demands.

As Malabou notes in *What Should We Do with Our Brain?*, although plasticity offers a possibility for conceptualizing the brain's historicity, in the sense that it becomes the account and the trace of how experience shapes itself into both matter and meaning, it persistently gets formulated as the mode and marker of a particular kind of individual identity, with history reduced to personal biography. Eric Kandel's widely read and cited neurobiological textbook coolly suggests that "it is this potential for plasticity of the relatively stereotyped units of the nervous system that endows us each with our individuality," while in Norman Doidge's popular account, *The Brain That Changes Itself*, neuroplasticity is explicitly framed as a marker of a pioneering individuality, with brains, as the back cover has it, demonstrating their potential to "repair themselves through the power of positive thinking."[6] The subtitle of the book, *Stories of Personal Triumph from the Frontiers of Brain Science*, makes clear the particular version of heroic individuality invoked: plasticity as liberal self-determination.

Malabou addresses this ideological insistence on plasticity as flexibility and adaptability in relation to capitalism's demands and as the marker of the experience of liberal, continuous, coherent individuality, with the question that underpins her book: "What should we do so that consciousness of the brain does not purely and simply coincide with the spirit of capitalism?"[7] The answer she offers is that we must awaken a historically conscious sense of plasticity: we must become aware of the historicity of neuronal matter itself and of the complex ways neurological form and function both shape and are themselves shaped according to experience with and within a material world. If the plastic brain both sculpts and is formed in relation to experience, she hints, a consciousness of these processes might become the marker and embodiment of a sense of history that is not limited to individuality's life story. Malabou also draws from her philosophical work the notion of a destructive plasticity that can explosively undo past formations,[8] finding in the possibility of a revolution within brain and psyche that appears following neurological damage something that fundamentally menaces the notion of the liberal autobiographical subject. For Malabou what seems really to matter about neuronal matter is indeed its potential to embody destructive plasticity: the possibility of a neurological revolution of the self, a negation without reserve, and a form of living beyond narrative continuity.

This notion of revolution that contains "the image of a world to come" seems at first glance to pull strongly against Luria's "romantic" "neurological novels" and the work of his strongest advocate in the United States and the United Kingdom, Oliver Sacks.[9] Malabou describes how both Sacks and his predecessor transform accounts of destructive plasticity into "epic narratives" of "heroes, victims, martyrs, warriors . . . travelers to unimaginable lands."[10] As Sacks himself put it in his introduction to *The Man with a Shattered World*, Luria's work was "always and centrally concerned with identity" and suffused with a "warmth, feeling and moral beauty"; it thus formed a line of continuity with his own liberal and humanist accounts of brain damage.[11] In such texts the self's connection with the past may be profoundly ruptured, but the narrative work undertaken by the author or clinician becomes, in Malabou's terms, a "clinical gesture" that enables these neurological case studies to become "mirrors in which we learn to look at ourselves." Malabou contends, "The problem of such case histories is how to do justice, in the very writing of the cases, to the rupture of narrativity that ultimately characterizes each one, to the destructive power of plasticity. . . . But what rhetoric could possibly account for the breakdown of connections, for destructive metamorphosis? And who would write the aphasic's novel? . . . *What mirror could reflect a brain*?"[12] She argues that to accede to narrative forms that cannot account for rupture and that overemphasize repair, even if just within the structure of humane meaning forged within the clinician's literary work, is to take the destructive plasticity of the brain wound and mold it according to a mode of reparation. Through such writing the clinician lends the brain-damaged subjects their meaning, their place within a synthesized narrative whole, but at the expense of both the phenomenological and political reality that might be glimpsed in destructive plasticity. Filling in the gaps with the plastic, moldable material of narrative marks a submission to what Malabou delineates in her earlier work as "the absence of revolution in our lives, the absence of revolution in our selves."[13] While attentive to many of their suggestions and possibilities, she finally finds in such "neurological novels" a reflection of the brain that supports a problematic naturalization effect and a conservative consonance with the way things currently are.

In order to take seriously Malabou's insistence on an account of the historicity of the plastic brain, how the brain is a "work" within which an immanent capacity for revolutionary reformulation is encoded and of which we are both authors and products, we would like, with due deference to destructive plasticity, to make a return to narrative, and to Luria and Zasetsky's narrative in particular. By giving back to Luria and Zasetsky's "neurological novel" its historicity—its place within the history of aphasiology, but also, more significantly, by restoring it to its Soviet context—by returning this *History of a Brain Wound* to itself, we hope to render explicit the entanglement of neurobiology with a revolutionary politics of the social that insists on the Soviet conception of plasticity. Contrary to Malabou's project, however, we are not suggesting the possibility of a political philosophy built on the idea of a given materiality of neuronal function that science is revealing to us. We are not claiming for Luria a discovery of the brain's essentially dialectical nature or a blueprint for a revolutionary subject built on the ontological primacy of an uncovered neurological substrate. Instead we want to demonstrate how Luria and Zasetsky's accounts of the plastic brain, and indeed the aphasic person embedded in a social world, are always and already objects and subjects of and in history in ways that demand the deconstruction of the idea of any given "brain fact" on which selves are built. The thoroughgoing historicity of Luria's account of neuroplasticity, and the narrative he shapes to express it, seem instead to model a self-conscious awareness of the historically embedded, politicized, neurological subject in which there can be no ontological separation between biological matter and the social contexts that mold it into meaningful formations.

Emily Martin has argued that the contemporary insistence on models of subjectivity that "begin with molecules, but go no farther than the central nervous system" has led to the emphasis on an "ahistorical concrete body" in neuroscience—a neurological self that measures its limits at the surface of the skin and that necessarily occludes the importance of social practices in accounts of its formation.[14] Against this "ahistorical concrete body" necessarily drained of the political, and even against that more yielding neuroplasticity that seemingly produces the potential for our individuality, we will use Luria's Soviet neurology to explore the formation of what might be termed a "historical plastic subject." For, despite its reception in the West, Luria's work always understood itself to be both political and historical in ways that usefully complicate the biological reductionism that reads the self as simply an epiphenomenon of neural architecture. By showing how minds and bodies extend into social practices and are shaped by those practices in return, Luria's work persistently suggests that the brain cannot adequately be understood as given or simply identical with itself; instead he conceptualizes it as dynamically implicated in a world-historical environment. As such, Luria's explicitly dialectical understanding of how brain matter forms itself inextricably from its experience of an extended, material world offers the possibility of a more historically and politically imprinted concept of plasticity than tends to be visible when it is absorbed under the master concept of identity. Rose and Abi-Rached have recently suggested that neural plasticity is one of the areas of contemporary biomedical science where there is an authentic and sophisticated "struggling toward a way of thinking in which our corporeality is in constant transaction with its milieu, and the biological and the social are not distinct but intertwined."[15] If this is so, perhaps Luria's history, alongside his historicity, can pry open space for an account of plasticity that appears less like a property of our being waiting to be found and more like a force of becoming that extrudes into visibility as it comes into contact with form, with historical materials. By tracking the idea that neuroplasticity might itself be understood as a kind of history—a history cut through with the potential for disaster but that is nevertheless informed by the possibility that things might be worked into another shape—we explore the emergence of a concept that takes the impression of how interior and exterior human worlds precisely become pieces of "work" that are mutually, dialectically constitutive of one another.

## From Destruction to Narrative Plasticity

THOUGH CONTEMPORARY NOTIONS of neuroplasticity emphasize flexibility, adaptability, and indeed healing after brain damage, the aphasia with which Zasetsky was diagnosed, as both an individual event and a historical object in the clinic, begins with destruction. The term refers to impairments of the expression and comprehension of language in any modality—in writing, speech, or use of linguistic signs—precipitated by acquired damage to the brain. The symptom complex was first definitively outlined in 1861 by Paul Broca in his famous presentation of a case of "aphemia" in a patient called Leborgne. Broca offered his case as part of a wider debate at the Société d'anthropologie in the Paris of the Second Empire (1852–70) on the question of "big heads" and whether there was a direct relation between brain size and intelligence. The debate soon centered on the question of language and whether even this highest of human faculties, usually associated with the immaterial parts of human nature, could be localized within specific areas of the material brain. Broca's belief in the principle of cerebral localization of faculties was demonstrated by the presentation of Leborgne's case. Paralyzed in the right leg and almost mute for twenty-one years, Leborgne had been able to make only the utterance *tan, tan*, alongside a few oaths and swear words. At autopsy his brain was discovered to have a lesion in the third frontal convolution of the left hemisphere caused by a cyst, and this area was thus put forward as the localized center for the "faculty of articulated language."

Joining clinical description with pathological anatomy, Broca, and those who immediately followed him, produced complex morphological studies of the structure of the brain and its "centers" upon which clinical models of the spatial localization of function were then built. Early classical aphasiology had little place for a concept of reparative plasticity, as its methods were based on the conception of static damage to centers that could be mapped at autopsy and then matched with the functional disturbances in speech that had been observed when the patient was alive. Consequently, as L. S. Jacyna has put it, and despite the efforts of various speech therapists, up until World War I the static diagram of damage to centers dominated aphasiology's imaginary, and a mode of "therapeutic nihilism" obtained.[16] But as Stefanos Geroulanos and Todd Meyers have demonstrated in this volume, the Great War was a world-historical event that precipitated a factory production line of relatively discrete head wounds that offered up a new kind of patient for aphasiology. Kurt Goldstein's *Aftereffects of Brain Injury in War* (1942) affirms that during the Great War doctors were no longer "dealing with progressive diseases, but with young men with long lives before them. Adaptation of the organism to the damage can be expected to some extent. The patients themselves show more willingness to learn, and there is much more hope of improving their mental condition, than in the case of diseased individuals."[17] And through the lens of a different kind of patient, the expression of a reparative neural plasticity thus began to extrude into visibility. For, though "defective" of speech, these men, who were frequently of the officer class, were not without language completely, and their education, class position, and masculinity rendered them trusted witnesses of their own disability.[18] By being possessed of a history their doctors could register, these were men who could be granted a legible position within an emerging medical narrative of rigorous examination, scientific testing, and then therapeutic progression.

As Geroulanos and Meyers show in their essay in this volume, Goldstein's systematized therapeutic regime emerged in tandem with a new theory of neurological organization—a theory that determined that brain-damaged patients would naturally compensate for their losses and reorient their brain functions toward a form of holistic order. Goldstein insisted that aphasia must be understood as a complex constellation of symptoms and physical compensations that were only triggered by an organic injury that could be observed and mapped onto a localized area of the brain. Indeed perhaps the most lasting contribution made by Goldstein to aphasiology was his assertion that even though there may be a physical lesion in the brain, aphasic symptoms always appeared as the result of integrations carried out by all available portions of the central nervous system—by what he called, in 1934, the "organism" as a whole. "Symptoms are answers, given by the modified organism, to definite demands," he writes.[19] Explicitly conceptualizing these integrations and reorderings through a notion of *Plastizität* as early as 1931,[20] Goldstein is clear, however, that the adaptive reaction to the experience of wounding does not represent a return to the track of the self that existed before the brain wound. Even in the early *Uber Aphasie* (1927) he is insistent: "Man is a psycho-physical organism. Each disease changes him in his entirety."[21]

Goldstein's account of aphasia differs greatly, then, from the static diagrams of the localizationists, with the temporally extended case study emerging in aphasiology as a mode capable of registering the paradoxical constellation of losses and "positive" symptoms and adaptations that track the complex progression of a nervous system reorienting itself in relation to both the internal and the external world. For a neuropsychological theory of language processing and recovery demands an emergent form—a form capable of following the complex torsions and vicissitudes of the expression of the relationship between neural matter, language, and thought as psychological phenomena. Taking account of patients' use of language and their subjective narratives rather than reducing them to diagrammatic representations also shifts clinical representations of aphasia from "therapeutic nihilism" to a new humanism—an empathetic engagement and the sense of a shared project of healing in which the clinician might hope to intervene therapeutically to become a protagonist, alongside the patient, in a tale of reintegration.

The temporally extended neurological case study is recentralized in the work of Goldstein, but it is brought into full narrative form by the tradition of "romantic" neuropsychology that Luria draws from him. Following Luria, and indeed following Sacks, however, narrated, even novelized case studies no longer seem at all unusual to us. Angela Woods has written of the privileged position of narrative and an idea of narrative collapsed into linear and therapeutic story that has come to dominate work in the medical humanities.[22] She convincingly argues for the need to think in more detail about how a concentration on particular kinds of storytelling found in the work of writers such as Arthur Frank—with an emphasis on linear narratives that stress deep psychological continuities across time and an expressive, confessional "I"—might privilege and render problematically universal modes of subjectivity and self-expression that are in fact highly culturally and historically contingent. Frank's emphasis on the importance of story in medicine, which emerged in the 1990s, argues assertively that the ill person is a "narrative wreck."[23] By this account patterned stories might assist modes of psychological reorientation toward meaningfully ordered experience, with linear narrative becoming a kind of prosthesis upon which the ill subject might lean in order to flesh out and fill in the disruptions and gaps in her individual embodied experience. But such a reification of certain modes of experience might fail to help us map the unexpected topographies encountered in disease and damage: the possibility of absolute psychic discontinuity; the common reality of chronic conditions that resist narratives of subjective overcoming; the inevitability of death, which can be seized as a pole of orientation but can never be a part of a subject's own narrative. As Malabou rightly notes, it is also the clinician, rather than the brain-damaged subject, who gets to shape the paradigm, the master plot of plastic reorganization.

Sacks certainly conceives of his own case histories, or what he calls "clinical tales," as narratives that strive to reorient the self in a world experienced as broken; they are concerned with "organisation and chaos, order and disorder."[24] A clinical tale aims to describe the relation of the patient's altered world to "our world." But what and whose world is "ours"? Notions of chaos and disorder rely on conceptions of disorganization and order, and such apparently stable entities are constantly shifting, contingent. Although Sacks, who was influenced by Goldstein's establishment of a deeply liberal, humanistic neurology in the United States after World War II, finds in *The Man with a Shattered World* an expression of "romantic" neurology that speaks in consonance with Western ideas of heroic, individualized overcoming, Sacks's world is not the same as Luria's. As is clear from his book *Traumatic Aphasia* (1947), Luria was significantly influenced by Goldstein's aphasiology and the need for a holistic analysis of any constellation of symptoms, though he was keen to emphasize his engagement with both "holist" and "classical" traditions in aphasiology.[25] But Sacks's emphasis on what Luria himself calls the "romantic" qualities of his "neurological novel" occludes the degree to which the worlds into which they were attempting to reintegrate their patients were ordered and understood as meaningful in definitively different ways. In what follows we hope to demonstrate that this difference in "world" becomes significantly visible in Luria's account of plasticity. And plasticity perhaps becomes particularly revealing in this regard because it seems to function in aphasiology as a concept possessed of a peculiarly *plastic* quality; it seems to find itself molded to the world from which it emerges. Part of the reason for this is that in the period preceding the emergence of direct rather than inferred evidence of neurogenesis, but even still to a large extent, plasticity occupies the territory of a concept rather than a "brain fact" in aphasiology. An entanglement of matter and idea, biology and culture, inner and outer demands, plasticity is a heuristic device that gets to stand in for various relatively poorly understood physiological mechanisms linked to psychological recovery. As such, plasticity registers the link between neuronal matter and correlated psychological functions, while having to bridge with its own moldable materials—materials susceptible to taking and holding the impression of the investments and attachments of individuals and cultures—the absence in understanding of a clear, causative relationship between the two.

## Plastic Materials

ON JUNE 22 1941, Germany broke its nonaggression pact with the Soviet Union. Operation Barbarossa was the largest invasion in military history. It was two weeks before Stalin finally addressed his Soviet "brothers and sisters," mobilizing the masses for "a war of the entire Soviet people" in which perhaps 20 million of the country's citizens would meet their deaths, the highest death toll of any nation involved in the conflict.[26] Like the October Revolution and the Stalinist "great break" that preceded it, World War II was a violently transformative moment in Soviet history. But the Great Patriotic War played a crucial function in reshaping the revolutionary narrative, disrupting and consolidating Soviet identities in new ways, and forging myths that would ultimately outlive the Soviet Union itself.[27] Echoes of this official state narrative can be found in Luria's own account of his scientific career. In his autobiography he writes:

> World War II was a disaster for all countries, and it was
> particularly devastating for the Soviet Union. Thousands of
> towns were destroyed, tens of thousands of people died from
> hunger alone. Many millions, both civilian and military, were
> killed. Among the wounded were thousands who suffered
> brain injury and who required extended painstaking care. The
> unity of purpose of the Soviet people so clearly felt during the
> great Revolution and the subsequent years re-emerged in new
> forms. A sense of common responsibility and common purpose
> gripped the country. Each of us knew we had an obligation to
> work together with our countrymen to meet the challenge. We
> each had to find our place in the struggle.[28]

Here we see Luria situating his medical work within the broader collective war effort, as he explicitly frames the experience of the conflict as part of the grand and ever-unfolding project of revolution. But if the Revolution represented a decisive rupture with all that came before it, then the war appears as a devastating external attack that united the nation around a shared mission to recover and restore what had been destroyed.

Not only did historical events have a direct impact on Luria's research, but the grand historical narratives of rupture and repair have a counterpart in his neurological work. Following the outbreak of war, Luria briefly joined the volunteer corps before being commissioned to organize a hospital to care for those requiring treatment away from the front. In the 1930s Luria had entered medical school and retrained as a neurologist, splitting his time between his studies in Moscow and psychological work in the Ukrainian city of Kharkov. He graduated from medical school in 1937,[29] and during the war the focus of his work became people who had survived traumatic brain injuries. He set up a four hundred–bed sanatorium in the southern Urals, overseeing the construction of laboratories and therapeutic training rooms along lines that mirrored Goldstein's institute. He worked there with a team of thirty researchers, whose tasks were to diagnose and treat brain injuries and to develop "rational, scientifically based techniques for the rehabilitation of destroyed functions." Luria notes that despite their modest equipment, their "most important resource was dedication to the task."[30] This was not routine scientific research but work carried out in service to the state and in the hope of ameliorating some of the damage inflicted by the most devastating conflict in human history.

Although Luria had begun his research into aphasic disorders in the 1930s, as with Goldstein before him, the war provided unprecedented opportunities for Soviet neurologists to examine the impact of localized brain injuries. The event arguably determined the trajectory of Luria's career, during the subsequent three decades of which he continued to probe the theoretical questions and forms of therapy that he began to explore with this mass of brain-injured patients. The priorities for treatment were collective and determinedly social; they were to rehabilitate people as swiftly as possible with the aim of returning them to service in the Red Army or, if their injuries were too severe, to a productive civilian life. The initial emphasis of treatment was on the healing of physical wounds. Following this stage Luria and his team focused on rehabilitating their patients, attempting to discover ways they might recover their lost functions. Like Goldstein's institute, their laboratories contained workshops, and therapeutic techniques incorporated writing and other practical tasks to allow people to relearn skills. This of course relied on an understanding of the brain as plastic, as capable of forging new connections in the wake of disaster.

Luria's *Traumatic Aphasia*, published in the USSR in 1947, draws explicitly on his wartime work. By 1943 he and his team had already amassed eight hundred case histories. In Luria's work the term *trauma* (*travma*), in accordance with its etymological roots, refers exclusively to physical wounds. In *Traumatic Aphasia* he gives little consideration to the emotional impact of war injuries or to the emotional efforts required to restore brain functions; instead trauma is something that happens to the tissue of the brain. Luria's use of the term *trauma* was consistent with the Soviet understanding of both the term and the experience of war in general, for the emotional impact of war was not something that the Red Army officially recognized or treated.[31] The state instead cultivated an image of brave masculine fighters seemingly impervious to the war's horrific events—figures whose individual identity and experience were subordinated to historical purpose. Catherine Merridale refers to "the disappearance of individual trauma as an issue of public debate" in the Soviet Union, tying it to the state denial of the hardship, suffering and terror it unleashed on the population in the 1930s, as well as a shifting emphasis from the individual to the collective.[32] Eyes needed to be dry and fixed resolutely on the bright future.

Luria's immediate postwar publications are concerned precisely with this future, figured through the possibility of recovering from serious cerebral damage—figured, in other words, through plasticity. Luria is clear, however, that brain cells do not simply regenerate. "The neuronal structures of the cortex, once destroyed, are incapable of regeneration," he writes. Total recovery is not a possibility, and an inert scar will always remain. Luria nevertheless notes "the high degree of plasticity [*plastichnost'*] shown by damaged functional systems, due to dynamic reorganization and adaptation to new circumstances." It *is* sometimes possible to overcome the impact of the injury, but through compensation, substitution, and reorganization within neuronal structures rather than any "regeneration and restoration of their morphological integrity."[33] Luria's research thus combines an acknowledgment of the permanence of injury with an insistence that patients could find new ways of performing the tasks previously undertaken by the damaged regions of their brain. As such, recovery becomes a process of discovery; subjects are oriented toward a new future in which there may be continuities with the past, but the past is repeated in a different way.

Because for Luria disruption caused by a brain injury is always understood in relation to the specific functions it precludes, he is never concerned with considering brains in isolation from human lives. Brain injury can be conceptualized only in relation to the patients' former existence and the regular flows of social life from which they suddenly find themselves wrenched: "Lesions exclude people from everyday life and work. Restoring their functions will enable them to reintegrate into society: How may such a patient be brought back into the daily round of social activity and work? What measures must be used so that this may be done as rationally as possible?" Brain injuries "disrupt the normal life of the patient, exclude him from social intercourse and from work, and may cause irreparable damage to his intellectual life." In Luria's work humans are precisely distinguished from animals through their greater social development: "Man's mental activity always takes place in a world of objects created during the development of society, is always directed towards them, and is frequently carried out with their aid." More developed levels of plasticity in humans also endow them with a greater ability to reorient themselves to the world, with the brain's capacity to find new ways of performing old tricks anchored in the patient's interactions with the external world. As a consequence Luria's research focuses on restorative training using tools as external mediators between the patient and the environment. Although this entails a long and arduous process, some progress can be made: "Human activity can be restored notwithstanding the irreversible changes affecting those areas of the cortex responsible for its performance."[34]

So there is in Luria's Soviet neuropsychology a persistent emphasis on the structural relationship between the external, social world and the functioning of any individual brain—a relationship of collectivity significantly mirrored at the neuronal level. One strong difference between Soviet plasticity and the Western model that has come to discursive prominence, then, is precisely the foregrounding of a collectivism in mental activity that forms part of a functional system "involving the participation of a group of concertedly working areas of the cortex."[35] Accordingly, for Luria a brain lesion does not simply destroy individual functions; it also disrupts a functional system. He determines a vital scientific rationale for localizing lesions (helping to reintroduce localization back into models of aphasia after Karl Lashley's 1929 "holist" theory of "equipotentiality" suggested that different cortical areas were functionally equivalent, and when one part of the brain is damaged, another can assume its function) through an emphasis on localizable "functional systems." Nevertheless by refusing what Juergen Tesak and Chris Code call the "two-dimensional connectionist view of the neoclassical model,"[36] Luria still argues for a temporal and "dynamic localization of function." His analysis of language as "a process model" provides space for strategies for rehabilitation precisely because his model is dynamic rather than static, with the brain conceptualized as a complex interactive system. There is an emphasis on the possibility of plastic reorientation but a reorientation within a determined system that foregrounds what is collectively concrete over any individual flexibility. Neural plasticity, by this account, works toward the fashioning of Soviet subjects who are imagined as both responding to the material experience of the world and capable of reshaping and seizing its forms.

In *Brain and History* (1979), Luciano Mecacci, who studied psychiatry in Moscow with Luria in the postwar years, suggests that the significance of Luria's psychological work is its materialist conception of the brain: "*The basic properties of the brain are plasticity and efficient functioning in its interaction with its environment. . . .* The brain is always considered in terms of its relationships with the global history of the organism and of the individual; it is viewed as an active and plastic instrument with which man grows and lives in a natural and social environment." And, like the brain itself, Luria's theories also "matured in the course of a dialectic interaction between exceptional historical, social, philosophical and scientific developments."[37] The war might have provided Luria and his team of researchers with an unprecedented number of brain-injured patients to observe and treat, but the theoretical approach taken to those patients had its origins in Luria's early psychological work, which developed in the two tumultuous decades following the October Revolution.

It is significant that Luria's autobiography does not begin with an account of his birth in 1902, but in 1917: "I began my career in the first years of the great Russian Revolution. This single, momentous event decisively influenced my life and that of everyone I knew."[38] His psychological work began in these heady and experimental postrevolutionary years, when a plethora of schools were jostling to elaborate an authentically Marxist approach to their subject. And in this preneurological work, *plasticity* (*plastichnost'*) also appears as a general term for understanding human development and the mutual interactions between people and their surroundings.[39] Luria's later understanding of the brain as an active organ needs to be understood as developing out of psychological research that similarly treated the whole human subject as both adaptable and adaptive. The "brain fact" of neuroplasticity in his work thus needs to be read within the context of postrevolutionary Soviet discourse.

David F. Hoffman discusses how the nascent Soviet state sought to inculcate new behavioral norms in its overwhelmingly peasant population in the prewar years, as it attempted to transform everyday practices and beliefs in order to create distinctly communist New Soviet People. Hoffman covers a range of Soviet initiatives focusing on literacy, cleanliness, and efficiency, all of which were in some sense directed toward "control of the living environment."[40] Crucially this civilizing mission relied on a particular understanding of the human that distinguished it from other modernizing regimes. In the words of Leon Trotsky, "The revolution gave a mighty historical impulse to the new Soviet generation. It cut them free at one blow from the conservative forms of life, and exposed to them the great secret—the first secret of the dialectic—that there is nothing unchanging on this earth, and that society is made of plastic materials."[41] According to Hoffman, 'It was [the Soviets'] belief in the plasticity of humankind that heightened their ambition to transform not only people's daily habits and culture but their modes of thinking and human qualities as well."[42] Though this process could be undeniably brutal, Hoffman argues that it relied on enlisting the active and enthusiastic participation of Soviet citizens rather than their passive, subjugated compliance.

Psychologists participated directly in formulating this understanding of human adaptability. In his overview of Soviet psychology published at the end of the Stalinist era, Raymond Bauer declares, "Of all questions of psychology, none has been more clearly saturated with political and social implications than that of the plasticity of the human organism, the question of whether human nature is set by man's biological endowment, or can be changed by environmental factors." According to Bauer, 1928 represented a pivotal moment in the development of Soviet psychology. That year saw the abandonment of the New Economic Policy that had reintroduced certain forms of private enterprise into the Soviet economy and the adoption of the First Five-Year Plan that entailed the collectivization of agriculture alongside a program of rapid industrialization. Bauer argues that a qualitatively different understanding of plasticity accompanied the onset of the First Five-Year Plan. An understanding of the subject as a passive, clay-like material, molded by external forces, was replaced by an emphasis on individuals' internal emotional energies and their capacity for dynamic intervention in the material world. Bauer discusses the paradoxical combination of this new theoretical emphasis on individual agency, with a concrete increase in political repression. For the "agency" of the ideal Stalinist subject was to be directed solely toward the construction of communism: "The assumption that man acts purposively to destroy the equilibrium between himself and his environment was the necessary premise for postulating that man has higher goals than the preservation of his own life, namely the building of socialism."[43] By placing the onus on the individual to cultivate an enthusiastic, self-sacrificing, and competitive commitment to communism, the Party was able to emphasize the subject's agency in the world, while exercising greater control.

Luria's work was attacked for cleaving too closely to the first of these two understandings of plasticity, for overemphasizing the impact of environment on psychology rather than highlighting the human capacity to shape and determine historical development.[44] But the politically charged denunciations of his work need not be taken at their own word, for Luria's theoretical approach did not conceive of the human as a passive entity responding to outside forces; instead, in conformity with the Stalinist values outlined by Hoffman and Bauer, it defined plasticity as a dialectical and purposive, mutually transformative relation between subject and world.

Luria's work in this period was explicitly connected to the Soviet project of creating new people and norms. In 1931 and 1932 he led two expeditions to Soviet Central Asia, where he hoped to trace the changes in thought that he assumed would accompany the rapid social changes occurring in the region, particularly the collectivization of agriculture and the Campaign to Eradicate Illiteracy. He defined the experiments as "a statement of the fundamental shifts that had occurred in human consciousness during a vigorous realignment of social history—the rapid realignment of a class society and a cultural upheaval creating hitherto unimagined perspectives for social development." The continuities with his later work are clear: he hoped to demonstrate the mutual interdependence of humans with their environment. But he intended not only to show that humans were formed by their historical circumstances; he aimed to prove that they were also capable of making changes in history: "Consciousness is the highest form of reflection of reality: it is, moreover, not given in advance, unchanging and passive, but shaped by activity and used by human beings to orient themselves to their environment, not only by adapting to conditions but in restructuring them."[45]

In this period Luria was also involved in Soviet child psychology (or "pedology"). Lisa Kirschenbaum discusses plasticity as an important concept for understanding Soviet conceptions of childhood, claiming that children formed "the pliable rising generation [that] could be moulded into the reliable vanguard of the revolution."[46] But this definition of plasticity misses the activist component of Soviet subjectivity described by Hoffman and Bauer. Luria's ideologically zealous colleague Aron Zalkind described children as an "extraordinary plastic material," more susceptible to change than their adult contemporaries, but this is not to be understood as straightforward indoctrination. For Zalkind children are required actively to intervene in society, forming miniature cadres of dedicated revolutionaries untarnished by the former "stagnant way of life," and setting an example to the older generation rather than meekly following its authoritarian lead.[47] Luria's own definition of plasticity in his work with children is also explicitly dialectical. In his early essay "Paths of Development through the Child" (1929) he defines "dialectical thinking" as the ability to "take into account all sorts of changing conditions. So as to be able not only to adapt to the real world but also to predict its dynamics and to adapt it to oneself." This, he claims, demands "a considerable plasticity and flexibility of behaviour that enable one to make use of different devices and different means, depending on the situation."[48] By the time Luria was to meet Zasetsky—a child formed between the October Revolution and the establishment of the USSR—he was able to match to the soldier's devastating brain damage and his historically and ideologically specific desires for recovery, concepts and narratives of neural plasticity, alongside therapies, that spoke not just to this damaged individual but to a devastated society that nevertheless held to the hope that another world might be possible.

### A World Shattered, a World Remade

ALTHOUGH THE MAN with a Shattered World is almost universally read in the West as inaugurating the genre of the "romantic" "neurological novel" that supports the reintegration and reformation of the liberal humanist individual, such an interpretation denudes this narrative of plasticity of its status as an object of history. Zasetsky, the case history's protagonist-patient, is certainly driven by a desperate desire to heal and to reintegrate into society, but the narrative's preoccupation with overcoming obstacles, of determined striving toward a final moment of reconciliation, is also definitively tied to its Soviet context. Indeed the text stages a dialectical tension between woundedness and healing, fragmentation and unity, that overwhelmingly conforms to the master plot of a socialist realist novel. Katerina Clark discusses the socialist realist novel as ritual—a cultural form and social act through which meaning and national myths were generated, consolidated, and sustained. The socialist realist master plot permeated and organized everyday discourse, and the genre proved remarkably consistent from its official inception at the 1934 Soviet Writers Congress up to the Brezhnev era, when Luria wrote his case history. Clark explains, "The formulaic signs of the Soviet novel . . . proved . . . tenacious over time. . . . The master plot . . . is the literary expression of the master categories that organise the entire culture."[49]

Unlike nineteenth-century realism, socialist realism was never intended as a mirror of the present; it was figured more like a window onto the future. From this tension between what is and what ought to be, the narrative derives its relentless forward-moving drive. The socialist realist novel is dynamic, with an emphasis on the ongoing struggle to shape and build communism, to bring a new reality into being, while the hero embodies the Marxist-Leninist account of history. The questing protagonist undergoes a transformation, which Clark characterizes as a working out of the dialectic between spontaneity and consciousness, the movement famously discussed by Lenin in his 1902 text *What Is to Be Done?* Over the course of the novel the hero masters himself, seizes his own form, and eventually achieves harmony with the movement of history.

*The Man with a Shattered World* at least attempts to trace a similar movement. The case history describes a heroic man—Luria explicitly describes him as such throughout[50]—struggling with almost superhuman effort and determination to overcome his own infirmity: "This book describes the damage done to a man's life by a bullet that penetrated his brain. Although he made every conceivable effort to recover his past, and thereby have some chance of a future, the odds were overwhelmingly against him. Yet I think there is a sense in which he may be said to have triumphed. The real author is its hero." But Zasetsky's acute sense of loss is not merely individual; it is also framed in terms of his sense of disaffiliation from society. He expresses regret that he can no longer "be of some service to my country" through work; nevertheless writing "the story of his life" becomes the means whereby he might become "useful to others."[51]

Nature often appears as the antagonist in socialist narratives of heroic struggle, with dam building, farm collectivizing, and factory constructing in hostile environments all featuring prominently. For Zasetsky, however, the nature to be overcome is not external but internal; he is at war with his own brain. And it is in its description of this psychic reality that Luria's narrative suggestively departs from the socialist realist master plot. According to Régine Robin, the hero of the socialist realist novel can undergo tragedy, can end up alone, can experience failure, conflict, and alienation, but he cannot, she insists, be uncertain of his destiny and the trajectory of history with which he is moving: "In the new man, there is necessarily and by definition . . . and in the midst of disasters, mud and horror—a wager on the future, on the construction of a new society, a wager on the well-foundedness of the struggle that reconciles him at some point . . . with the movement of history, with which he turns out to be in harmony. . . . There is the idea . . . that the deficiencies will be overcome, and that the transparency of social and individual relations will finally be established."[52] Zasetsky himself does not express doubt about the well-foundedness of the struggle he continues to wage; nevertheless, although ostensibly staging a transition from spontaneity to consciousness, the work is not, finally, a portrait of reconciliation. The pieces cannot be reassembled, molded into a coherent totality. Though the novel is propelled by the unifying zeal of Soviet discourse, the neurological condition being described resists the final glimpse of totality necessitated by socialist realist narratives. Instead there is perhaps a more negative dialectic at work that ensures the narrative never achieves the smooth continuity of socialist realism in which aesthetic form synthesizes content into a coherent present.

"Who would write the aphasic's novel?" Malabou asks. She suggests that instead of a Luria or a Sacks, Samuel Beckett, with his modernist aesthetic of rupture and exhaustion, with his rhetoric comprising "figures of interruption, pauses, caesuras—the blank spaces that emerge when the network of connections is shredded," might offer a mimesis consonant with the world of the "new wounded."[53] But on close reading Luria's "neurological novel" is also significantly gashed and gapped in a way that holds it distinctively apart from Sacks's narratives. For, despite all his strenuous efforts, Zasetsky cannot retell a story, even a child's fable—"disjointed phrases and clauses were all that occurred to him"—and although Luria details and quotes from the almost three thousand pages Zasetsky wrote in an attempt to unify his experience and compensate for his losses, the doctor resists the urge to form the book into a unified totality. Although Zasetsky states, "This writing is my only way of thinking. If I shut these notebooks, give it up, I'll be right back in the desert, in that 'know-nothing' world of emptiness and amnesia," Luria is insistent that despite this heroic work, coherence cannot, will not be restored: "He desperately wanted to wake from this terrible dream . . . to find the world clear and comprehensible instead of having to grope for every word he uttered. But it was impossible."[54]

Luria preserves in the text surreal images that might indeed come from the pages of Beckett: spoons appear as frightening "bits of space"; once familiar towns are completely bewildering; heads become as large as tables one minute and as small as chicken heads the next; hands and legs vanish and must be "hunted"; written language appears as a mysterious series of squiggles. Zasetsky sees everything as though through a swarm of tiny flies, with words too appearing as though they had been "gnawed, plucked around the edges, and what's left are scattered points, quills, or threads that flicker like a swarm." The narrative is also persistently rent with elisions and lacunae, particularly in the early sections, which utter in a voice of frustrated inability, "How horrible this illness is! I still can't get a grip of myself, can't work out what I was like before, what's happened to me."[55]

Zasetsky's wounding itself appears as an absence, an ineffable caesura in his biography: "Under fire, I jumped up from the ice, pushed one . . . towards the west . . . there . . . and . . . Somewhere not far from our furthest position on the front lines, in a tent blazing with light, I finally came to again." The translator of the book, Lynn Solotaroff, notes that "Luria has scrupulously preserved the repetitions and inconsistencies which are symptoms of the patient's condition" within the torn fabric of his account; but far more striking than this local detail is the way that Zasetsky's and Luria's interlaced and sometimes peculiarly merged narratives persistently pull against the clarity of any broader synthesizing mode. For it seems initially as if doctor and patient are set clearly alongside one another within Luria's narrative frame. Luria (who is listed as the author on the title page) speaks coherently in the past tense, while Zasetsky frequently slips into the present tense, as if unable to steel a grip on the past, and utters in a voice characterized by its affective intensity and abrupt, frustrated elisions. But as the narrative progresses Luria begins to slip into a clipped and cryptic style that seems to mirror his subject. "In the beginning it was all so simple. His past was much like other people's: life had its problems, but was simple enough, and the future seemed promising," he tells us. "Then suddenly it was all over."[56] And although Luria may be in control of the overarching structure, it is Zasetsky's name that sits under the section entitled "From the Author," and a large portion of his writings are reproduced seemingly verbatim. The shift from Luria to Zasetsky in the narrative is never formally signaled, and even when Luria presents a number of lengthy scientific asides ("digressions," he calls them suggestively), which read more clearly as case notes and case history, he soon regresses, in an oddly seamless fashion, back into the "neurological novel" style of quotation from Zasetsky and contextualizing material. Here clinical discourse does not maintain its distinctness, its framing difference, from the language of aphasic struggle.

"What matter who's speaking," wrote Beckett in *Texts for Nothing*; it matters here simply in the sense that the ambiguity signals how Luria allows Zasetsky's words, his aphasia, to leak out of a containing frame. In a way that perhaps emphasizes commonality or a transferable shared place in a wider system, Luria even gives space to Zasetsky's thought experiments that demonstrate his "marked capacity for fantasy and empathy." He imagines living other lives, and one of these fantasies is particularly suggestive of a sense of collective endeavor that seems to have been present on both sides of the doctor/patient divide. Luria speaks of Zasetsky writing "with the precision of someone doing psychological research—someone who really knows his field," while Zasetsky projects himself into Luria's place: "Say I'm a doctor examining a patient who is seriously ill. I'm terribly worried about him, grieve for him with all my heart. (After all, he's human too, and helpless. I might become ill and also need help. But right now it's him I'm worried about—I'm the sort of person who can't help caring.)"[57]

The epilogue also has something of this odd, uncannily merged quality. "Were It Not for War . . . " presumably belongs to Zasetsky, though no author is signaled. It does repeat Zasetsky's earlier denunciation of oppression, slavery, and war and also extends a thought from his earlier "From the Author" piece about "the cosmos and outer space,"[58] with its dream of interplanetary travel and opportunities to access new natural resources imagined as infinite. But the lack of identifying signature extends the epilogue beyond the patient's personal statement and seems to speak in both collective and concrete terms. Luria writes in his chapter "Romantic Science" in *The Making of Mind* of his admiration for Lenin's observation "that a glass, as an object of science, can be understood only when viewed from many perspectives. . . . And the more we preserve the whole wealth of its qualities, the closer we come to the inner laws that determine its existence. It was this perspective that led Karl Marx to describe the process of scientific description with the strange-sounding expression 'ascending to the concrete.'"[59] If specificity and subjectivity are key, but only as paths to what will finally be synthesized as concrete, the individuality of the doctor, alongside the subjective experience of the patient, also become important only insofar as they illuminate a collective, shared, and objective reality.

In its shattered qualities, then, the text does not quite conform to Malabou's account of the case history as a "narrative intrigue" in which the clinician usurps the voice of the patient.[60] Instead the scarred surface of the case history captures a tension between woundedness and healing that retains hope for the future. The narrative clatters between despair and the hope, between the "I've never been able to put my life together again" of how things are and the "I hoped I'd be able to tell people about this illness and overcome it" of a potential future synthesis. The contradiction even expresses itself in the movement between the initial title of "The Story of a Terrible Brain Injury" and Zasetsky's final, preferred "I'll Fight On," in English "The Man with a Shattered World" and in Russian "A World Shattered and Remade." In the face of the present impossibility of synthesis, Luria concludes the book by describing how Zasetsky faces the future through continuing struggle, perhaps even a mode of permanent revolution: "He continues to try to recover what was irretrievable, to make something comprehensible out of all the bits and pieces that remain of his life. He has returned to his story and is still working on it. It has no end."[61]

Malabou finds in her concept of destructive plasticity a refusal of "the promise, belief, symbolic constitution of all resources to come"; it operates instead in the open space left in the wake of the "collapse of messianic structures." Destructive plasticity refuses salvation and redemption, placing a prohibition on thinking other possibilities: "It has nothing to do with the tenacious, incurable desire to transform what has taken place." Instead, she insists, destructive plasticity "deploys its work starting from the exhaustion of all possibilities . . . when cohesion is destroyed, family spirit vanished, friendship lost, links dissipated in the ever more intense cold of a barren life."[62] But in the content and the unexpected form of *The Man with a Shattered World*, one finds a rather different account of destruction. Luria does bear witness to destructive plasticity in this text: the cortex cannot regenerate; Zasetsky knows he will not be returned to wholeness. Things remain shattered, disconnected, in pieces. And yet from this woundedness something else also emerges: a pole of orientation toward a future that would not be a rebuilding of what went before but the possibility of another world. Indeed perhaps the very choice of Zasetsky as an exemplar of an aphasia is significant because, *pace* Malabou's emphasis on the revolutionary importance of the loss of affect, of coolness and indifference—of absolute change—Luria finds in Zasetsky the expression of a contradiction between irreparable destruction and the cognizance of his losses that leads to the hope of another kind of future. Luria writes that Zasetsky "is robbed of any possibility of a future and loses precisely what it is that makes a person human," and yet "the frontal cortex, had been spared, and with [it] his capacity to recognize his defects and wish to overcome them. He was acutely aware of what it means to be human and, to the extent his strength permitted, worked feverishly to overcome his problems. He suffered intensely."[63] Zasetsky did recover some capacity, though there was no return to a previous self. He learned to write again using his habit body—the motor automatisms laid down in the plastic nervous system that were preserved in his injury and that he used to support his other damaged functional systems. He positioned himself toward the future even though he could no longer conceptualize it. What is materialized in Zasetsky's brain, then, is precisely a dialectic of destructive and reparative plasticity that orients itself both from and toward a changed world but refuses to synthesize itself into the form of what Theodor Adorno would call an unreconciled totality. Writing from inside the impossible, paradoxical conditions of the Soviet real, narrated plasticity thus becomes something like a synecdoche for Luria—the molded image, the place, and the time of a wounded world that nevertheless holds within itself the possibility of a future.

Read in this way *The Man with a Shattered World* becomes simultaneously a narrative of loss and anticipation. If, as Clark maintains, socialist realism is a form of ritual, then Zasetsky's attempts to describe his life might be understood in similar terms: the beginning is returned to again and again. Returning to the site of disaster gestures toward the possibility of starting over or at least toward the hope of recovering something that was lost. Significantly, then, when understood in relation to the collective revolutionary history to which Luria and Zasetsky refer and in which both are embedded, the text's formal oscillation between Soviet reconciliation and modernist fragmentation tacitly acknowledges the failures of state socialism to establish a reconciled whole in which the wounded subject might ultimately be healed. Zasetsky and Luria indeed share a shattered world; nevertheless they still hope that the past might be redeemed. The future they envisage does not imagine that ruins can be reconstructed into the forms they once held; instead they hope for another world, a different world, which could shape itself in relation to the destruction of the past and present. In Luria's representation plasticity thus takes the shape of the "how it is" but holds it in a dialectical, as yet unreconciled relation to the "how it ought to be." The concept of plasticity seems to tread the boundary between destruction and repair in a way that might allow us to see our neural biology as always already forming itself in relation to, while also simultaneously constituting, human history and culture. But Luria's specifically Soviet plasticity also lends a visible shape to the possibility of a neurological subject who can fashion a social future that could be different to what has come before. This plasticity, understood as a dialectic, might offer some powerful tools for thinking through the knotted, entangled, historical relationship between what is given and what might be able to be seized.

# Endnotes

1  A. R. Luria, *The Man with a Shattered World*, trans. Lynn Solotaroff (Harmondsworth, UK: Penguin, 1975), 47. Unless otherwise noted, subsequent references cite this edition.

2  Catherine Malabou, *The New Wounded: From Neurosis to Brain Damage*, trans. Steven Miller (New York: Fordham University Press, 2012), 57; Luria, *The Man with a Shattered World*, 24.

3  Donald Hebb, *The Organization of Behavior: A Neuropsychological Theory* (New York: Wiley, 1949), 70.

4  Nikolas Rose and Joelle M. Abi-Rached, *Neuro: The New Brain Sciences and Management of the Mind* (Princeton, NJ: Princeton University Press, 2013), 52.

5  Jan Slaby and Suparna Choudhury, "Proposal for a Critical Neuroscience," in *Critical Neuroscience: A Handbook of Social and Cultural Contexts of Neuroscience*, ed. Suparna Choudhury and Jan Slaby (Oxford: Blackwell, 2012), 29.

6  Eric R. Kandel, James H. Schwartz, and Thomas M. Jessell, *Principles of Neural Science*, 4th edition (New York: McGraw Hill, 2000), 34; Norman Doidge, *The Brain That Changes Itself: Stories of Personal Triumph from the Frontiers of Brain Science* (London: Penguin, 2007).

7  Catherine Malabou, *What Should We Do with Our Brain?*, trans. Sebastian Rand (New York: Fordham University Press, 2008), 12.

8  See, Catherine Malabou, *The Future of Hegel: Plasticity, Temporality, Dialectic* (London: Routledge, 2004).

9  Malabou, *What Should We Do with Our Brain?*, 82.

10  Malabou, *The New Wounded*, 55.

11  Oliver Sacks, foreword to A. R. Luria, *The Man with a Shattered World*, trans. Lynn Solotaroff (Cambridge, MA: Harvard University Press, 1987), xvi.

12  Malabou, *The New Wounded*, 54, 55.

13  Malabou, *What Should We Do with Our Brain?*, 66.

14  Emily Martin, "AES Presidential Address: Mind-Body Problems," *American Ethnologist* 27, no. 3 (2000): 574, 581.

15  Rose and Abi-Rached, *Neuro*, 3.

16  L. S. Jacyna, *Lost Words: Narratives of Language and the Brain* (Princeton, NJ: Princeton University Press, 2000), 206.

17  Kurt Goldstein, *Aftereffects of Brain Injuries in War* (New York: Grune & Stratton, 1948), 66.

18 See Jacyna, *Lost Words*, 146–70, 204–30.

19 Kurt Goldstein, *The Organism: A Holistic Approach to Biology Derived from Pathological Data in Man* (New York: Zone, 1995), 35.

20 See Kurt Goldstein, "Über die Plastizität des Organismus auf Grund von Erfahrungen am nervenkranken Menschen," in *Handbuch der normalen und pathologischen Physiologie*, vol. 15, ed. A. Bethe (Berlin: Springer, 1931), 1133–74.

21 Kurt Goldstein, *Selected Papers/Augewählte Schriften* (The Hague: Nijhoff, 1971), 140.

22 Angela Woods, "The Limits of Narrative: Provocations for the Medical Humanities," *Medical Humanities* 37, no. 3 (2011): 73–78.

23 Arthur Frank, *The Wounded Story-teller* (Chicago: University of Chicago Press, 1995), 55.

24 Oliver Sacks, "Clinical Tales," *Literature and Medicine* 5 (1986): 18.

25 Luria did emphasize a "holist" refusal of the "reduction of living reality with all its richness of detail to abstract schemas," but he always held this emphasis on reintegration and plastic adaptation in a dialectical tension with his commitment to classical theories of brain localization. See A. R. Luria, *The Making of Mind: A Personal Account of Soviet Psychology* (Cambridge, MA: Harvard University Press, 1979), 174. Luria writes, "While, on the one hand, the mechanistic view of strict localization always leads the analysis of the cerebral basis of mental activity into an impasse, the holistic . . . opinions of mental processes were unable, on the other hand, to create the necessary preconditions for scientific progress." Quoted in Juergen Tesak and Chris Code, *Milestones in the History of Aphasia: Theories and Protagonists* (Hove, UK: Psychology Press, 2008), 155.

26 Richard Overy, *Russia's War: A History of the Soviet War Effort 1941–1945* (London: Penguin, 1997), 79.

27 See, for example, Amir Weiner, *Making Sense of the Second World War and the Fate of the Bolshevik Revolution* (Princeton, NJ: Princeton University Press, 2001).

28 Luria, *The Making of Mind*, 138.

29 See, for example, E. D. Homskaya, *Alexander Romanovich Luria: A Scientific Biography*, trans. David E. Tupper (New York: Kluwer Academic/Plenum, 2001); Elena Luria, *Moi Otets A. R. Luria* (My Father A. R. Luria) (Moscow: Gnosis, 1994).

30 Luria, *The Making of Mind*, 139.

31 Evidence of neuroses unleashed by the siege of Leningrad recorded by doctors at the time, for example, remain buried in archives, as they contradict the official heroic narrative. See Catherine Merridale, "The

Collective Mind: Trauma and Shell-Shock in Twentieth-Century Russia," *Journal of Contemporary History* 35, no. 1 (2000): 48.

32  Ibid., 46.

33  A. R. Luria, *Restoration of Function after Brain Injury*, trans. Oliver Zangwill (London: Pergamon Press, 1963), 32, 33.

34  Ibid., 259, xi, 39, 154.

35  A. R. Luria, *The Working Brain: An Introduction to Neuropsychology* (Harmondsworth, UK: Penguin, 1973), 35.

36  Tesak and Code, *Milestones in the History of Aphasia*, 158.

37  Luciano Mecacci, *Brain and History: The Relationship between Neurophysiology and Psychology in the Soviet Union*, trans. Jenry A. Buchtel (New York: Brunner/Mazel, 1979), xxiii, xxii.

38  Luria, *The Making of Mind*, 3.

39  See, for example, L. S. Vygotsky and A. R. Luria, *Etyudi po Istorii Povedeniya: Obez'yana, Primitiv, Pebyonok* (Moscow: Gosudastvennoe Izdatel'stvo, 1930), 130; L. S. Vygotsky and A. R. Luria, *Studies on the History of Behaviour: Ape, Primitive and Child*, ed. and trans. Victor I. Golod and Jane E. Knox (Hillsdale, NJ: Lawrence Erlbaum, 1993), 175.

40  David F. Hoffman, *Stalinist Values: The Cultural Norms of Soviet Modernity, 1917–1941* (Ithaca, NY: Cornell University Press, 2003), 45.

41  Leon Trotsky, *The Revolution Betrayed: What Is the Soviet Union and Where Is It Going?*, trans. Max Eastman (Mineola, NY: Dover, 2004), 120.

42  Hoffman, *Stalinist Values*, 45.

43  Raymond A. Bauer, *The New Man in Soviet Psychology* (Cambridge, MA: Harvard University Press, 1952), 79, 80.

44  See David Jorawsky, *Russian Psychology: A Critical History* (Cambridge, MA: Basil Blackwell, 1989), 247.

45  A. R. Luria, *Cognitive Development: Its Cultural and Social Foundations*, trans. Martin Lopez-Morillas and Lynn Solotaroff (Cambridge, MA: Harvard University Press, 1976), 19, 8.

46  Lisa A. Kirschenbaum, *Small Comrades: Revolutionizing Childhood in Soviet Russia, 1917–1932* (New York: Routledge/Falmer, 2000), 163.

47  Aron Zalkind, "The Pioneer Youth Movement as a form of Cultural Work among the Proletariat," in *Bolshevik Visions: The First Phase of the Cultural Revolution in Soviet Russia*, ed. William G Rosenburg (Ann Arbor, MI: Ardis, 1984), 351, first published in Russian in *Vestniktruda* 3, no. 40 (1924): 107–16.

48  A. R. Luria, "Paths of Development of Thought in the Child," in *Selected Writings of Alexander Luria*, ed. Michael Cole (White Plains, NY: M. E. Sharpe, 1978), 99.

49   Katerina Clark, *The Soviet Novel: History as Ritual* (Chicago: University of Chicago Press, 1981), 14.

50   This is even more pronounced in the Russian text: certain passages, which underline and repeat the militaristic tropes, are omitted from the English translation altogether. Compare, for example, p. 105 in the Russian with p. xix in the English, or p. 17 in the English with p. 117 in the Russian. A. R. Luria, "Poteryannii i vozvrashchennnyi mir," in *Romanticheskie Esse* (Moscow: Pedagogika Press, 1996).

51   Luria, *The Man with a Shattered World*, 16, 35, 84.

52   Régine Robin, *Socialist Realism: An Impossible Aesthetic*, trans. Catherine Porter (Palo Alto, CA: Stanford University Press, 1992), 242–43.

53   Malabou, *The New Wounded*, 55.

54   Luria, *The Man with a Shattered World*, 72, 78, 128.

55   Ibid., 38, 26.

56   Ibid., 22, 25, 21.

57   Ibid., 127, 82, 127.

58   Ibid., 18.

59   Michael Cole, Karl Levitin, and Alexander Luria, *The Autobiography of Alexander Luria: A Dialogue with The Making of Mind* (Mahwah, NJ: Lawrence Erlbaum, 2006), 177–78.

60   Malabou, *The New Wounded*, 55.

61   Luria, *The Man with a Shattered World*, 116, 77, 71, 129.

62   Catherine Malabou, *The Ontology of the Accident: An Essay in Destructive Plasticity*, trans. Carolyn Shread (London: Polity Press, 2012), 89–90.

63   Luria, *The Man with a Shattered World*, 71.

David Bates

# 7 Automaticity, Plasticity, and the Deviant Origins of Artificial Intelligence

THE CONTEMPORARY BRAIN is largely a digital brain.[1] Not only do we study the brain through virtual technologies that rely on digital visualizations, but the brain's very activity is often modeled by a digital simulation.[2] And the brain is, in many different ways, still understood to be a digital machine; it is a kind of neural computer.[3] The legacy of artificial intelligence (AI) still persists in contemporary neuroscience and cognitive science.[4] The two competing projects to "map" the brain in Europe and the United States clearly reveal the discursive and conceptual connections between computers and neurophysiology—and *neuropathology.* As the European Union Human Brain Project website puts it, understanding the human brain will allow us to "gain profound insight into what makes us human, develop new treatments for brain disease and build revolutionary new computing technologies."[5] Similarly the Connectome project initiated by the National Institutes of Health and supported by President Obama with substantial funding, reveals the essential link between new research technologies, the technologized vision of the brain itself, and pathologies: "Long desired by researchers seeking new ways to treat, cure, and even prevent brain disorders, this picture will fill major gaps in our current knowledge and provide unprecedented opportunities for exploring exactly how the brain enables the human body to record, process, utilize, store, and retrieve vast quantities of information, all at the speed of thought."[6]

It would seem that the recent intensification of interest in the inherent plasticity of the brain—its developmental openness, its always evolving structure in the adult phase, and its often startling ability to reorganize itself after significant trauma—puts considerable pressure on the technological conceptualizations of the brain that assume a complex but definite automaticity of operation. Indeed the concept of plasticity has been heralded as a counter to the machinic understanding of the brain, most notably by the philosopher Catherine Malabou.[7] However, it is now the case that the neurophysiological phenomenon of brain plasticity is rapidly becoming assimilated to computational models and digital representations of the nervous system. In the field of computational developmental neuroscience, for example, the brain is figured as a *learning machine* that automatically constructs, according to set algorithms, connective webs that are dependent on the specific "experience" of the network. These models are all ultimately derived from the seminal theory of the neurophysiologist Donald Hebb. As he famously explained in his 1949 book *The Organization of Behavior*, synaptic connections between neurons are strengthened with use; the theory is often reduced to the paraphrase "Neurons that fire together wire together." The processes governing the determination of the plastic brain as it experiences the world are obviously much more complex, but the basic principle still holds. Therefore even the contingent determination of the plastic brain can, it is thought, be rigorously modeled by a virtual computer simulation.

This doubling of the brain by its digital other has in turn affected the technological domain of computing itself. Attempts to model the cortex of animal brains with "synaptic" architectures, for example, are framed as experimental investigations of the neural organization itself; it is said that this is "a significant step towards unravelling the mystery of what the brain computes," which in turn, the researchers claim, open "the path to low-power, compact neuromorphic and synaptronic computers of tomorrow."[8] Digital visions of the plastic brain have stimulated the invention of new computational architectures. IBM, for example, has designed a chip that mimics the networked connectivity of neural systems. The chip has "digital neurons" that can rewire synapses based on their inputs.[9] A rival Intel project also promises programmable "synapses" and the imitation of "integrating-neurons."[10] These neurosynaptic chips are part of a broader investigation of what is being called "cognitive computing," whose aim "is to develop a coherent, unified, universal mechanism inspired by the mind's capabilities." That is, the researchers "seek nothing less than to discover, demonstrate, and deliver the core algorithms of the brain and gain a deep scientific understanding of how the mind perceives, thinks, and acts." Here the brain has become an algorithmic learning *machine*; it is a mirror of the very technology that represents it. Analysis of the brain's networked organization will, it is claimed, "lead to novel cognitive systems, computing architectures, programming paradigms, practical applications, and intelligent business machines."[11]

This mirroring has in fact a long history—arguably as long as the history of computing itself. The goal of early AI research was twofold: to produce an intelligent simulation and, by doing so, test hypotheses about the functions of the human mind. Underlying this project was the assumption that the mind or brain would be amenable to such analysis. As researchers wrote in 1954, with respect to their own effort at cortical modeling, "No matter how complex the organization of a system . . . it can always be simulated as closely as desired by a digital computer as long as its rules of organization are known."[12] This was, they note, the implication of Alan Turing's early conceptualization of the universal digital computer as a machine that can (virtually) imitate any other discrete-state machine.

The automaticity of the brain's operation (even in its most radically plastic guise) is to a great extent a consequence of the historical codevelopment of computer technologies, AI, and the cognitive sciences. The question of *autonomy* in this framework of automaticity is drained of all potential meaning. And so, once plasticity has been fully integrated into the computational and neurophysiological models of the brain, resistance to the total automatization of human thinking cannot simply rely on romanticized concepts of selfhood or philosophical critiques of materialism; we must focus instead on the historical and conceptual foundations of the digital brain itself. Resistance can be generated, that is, through a critical history of *automaticity*. By returning to the threshold of the digital age, that moment when the modern computer was first being conceptualized (and the ideas and practices of AI simultaneously set in motion), we can see that the digital was not, at the beginning, fully aligned with automaticity. Indeed, although it has not received much attention, key developers of early computer technologies were explicitly trying to imitate, with their new thinking technologies, a more radical openness, a more unpredictable plasticity within the nervous system—a subject that was, we will see, very much alive in early twentieth-century neurophysiology and neurologically informed experimental psychology. At the same time that some cyberneticians were claiming that the brain was just an automatic calculator like the computer, crucial figures in the history of computing and cybernetics immediately recognized the importance of the plasticity of the brain for the project of AI: the plastic brain, it was thought, offered the possibility of modeling creative, unpredictable leaps of human intelligence, capacities that went *beyond* the relentlessly automatic performance of rigid functional mechanisms or habitual behaviors. It is therefore significant that the neurophysiological discourses of plasticity in the period were intimately linked to the disorders and crises of the diseased or injured brain. Constructing a plastic computational machine at the dawn of the digital age therefore entailed, I will argue, the invention of a *machinic pathology*.

Recuperating this historical moment will offer a new perspective on our contemporary "digital brain." We need not be reduced to mere learning machines, largely unconscious of our own cognitive processes, where any experience of freedom and contingency of thinking can be exposed as some kind of Nietzschean illusion. The human brain was understood to be a special kind of genuinely open system that determined itself yet was also capable of defying its own automaticity in acts of genuine creativity. The originators of the digital computer were explicitly inspired by this neurophysiological concept of plasticity in their efforts to model the abilities of truly intelligent beings.

HISTORIES OF AI usually trace the lineage back through figures who attempted to simulate thinking in some kind of automatic machinery. It could be said that René Descartes's philosophical and physiological writings gave us a vision of the first modern automaton—that is, a *thinking* machine. While he of course resisted the ultimate implications of his own systematic mechanization of the human and animal body, Descartes pointed the way to a mechanical understanding of cognition when he gave a comprehensive description of the nervous system and the essential role it played not just in governing all animal behavior but also, more importantly, in producing the vast majority of *human* thinking and action, the routine cognition of the everyday.

Yet a closer look at Descartes's writing reveals a theory that is somewhat more complex than a merely mechanical vision of the body's operations. Descartes showed how the nerves and the brain were an information system, remarkable for its flexibility and adaptability. While the nervous system was a material structure, it was nonetheless plastic and modifiable. As a space for integrating information, the brain was, he wrote, "soft and pliant" ("molle et pliante") and capable therefore of being imprinted with memories and of acquiring reflexes.[13] The implication was that the cultural determination of the individual—through language, culture, history—took place within the soft architecture of the brain itself. As he remarked in the *Discourse on Method*, "I thought too, how the same man, with the same mind, if brought up from infancy among the French or Germans, develops otherwise than he would if he had always lived among the Chinese or cannibals."[14] The open brain was determined by the physical flows that were produced by the organs of sense and transmitted through the conduits of the nerves. Descartes's automaton was no mere clockwork mechanism but an open site of perpetual organization and reorganization occasioned by information received through the sensory systems.[15]

It is true that the long history of automated thinking technologies shows that the idea of a "soft and pliant" system was overshadowed by the hard mechanisms of the industrial age. From the semi-automatic calculators of Pascal and Leibniz to Charles Babbage's Difference Engine (the first example of a truly automatic form of artificial thought) and the unfinished general computer he called the Analytic Engine, and then on to the "logic pianos" of the late nineteenth century (built by, for example, William Jevons and Allan Marquand), models of human reasoning were concretely instantiated by machines with precise and predictable operations. It is hardly surprising, then, that these early examples of artificially mechanized reasoning would greatly influence the conceptualization of human cognition. For example, Charles Sanders Peirce, the American pragmatist philosopher, reflecting on logic machines, noted that the performance of a deductive rational inference had "no relation to the circumstance that the machine happens to work by geared wheels, while a man happens to work by an ill-understood arrangement of brain-cells."[16]

Peirce himself had the insight (many years before Claude Shannon) that these logical relations could even be performed by means of electrical circuits.[17] With this step we are on the very edge of modern computing and, with it, the effort to understand the brain itself as a machine for thinking, constructed from so many neural "switches."

We should pause, however, to remark that in the period before digital computing, knowledge of the brain was not at all congruent with the mechanistic paradigm of automatism. By the late nineteenth century the idea that the brain was made up of multiple, localizable components was giving way to models of the brain that emphasized its distributed structural organization. The great British neurologist John Hughlings Jackson, for example, used extensive clinical investigation of neural disorders to argue that the brain had many levels of organization due to the gradual development of the organ in the course of evolution. What is interesting is that Hughlings Jackson wanted to demonstrate that the highest intellectual capacities were associated with the *least* organized, least "evolved," and what he called "least automatic" cortical structures. The most automatic functions governed our basic physiological systems, while the relatively open and undetermined cortex was associated with complex thinking and discursive language.

William James, perhaps the most influential figure in the synthesis of psychology and neuroscience at the turn of the century, was very much interested in how the human mind was driven by habitual and unconscious cognitions. These were not given but rather acquired. The brain was at once a site of openness and a space of artificial mechanisms: "The phenomena of habit in living beings are due to the plasticity of the organic materials of which their bodies are composed." "Plasticity," wrote James, "means the possession of a structure weak enough to yield to an influence, but strong enough not to yield all at once." For James, as for Peirce, the organismic forms of life were essentially plastic since they were the product of evolutionary change, and the nervous system was the privileged site for this formation and reformation as the organism adapted to its changing environmental conditions. "Organic matter, especially nervous tissue, seems endowed with a very extraordinary degree of plasticity," he noted, and this explained the way beings could become automatons of a sort—but never true automata, precisely because plasticity never entirely disappeared from the brain and nervous system.[18]

Drawing on the evolutionary theories of figures such as Herbert Spencer and Hughlings Jackson, James took up the idea that the human capacity for intuition and insight was most likely linked not to some specific organ of intelligence but in fact to the instability and indetermination of the higher centers of the brain. James suggested that "the very vagueness [of the cerebral hemispheres] constitutes their advantage. . . . An organ swayed by slight impressions is an organ whose natural state is one of unstable equilibrium." This "plasticity" of the higher brain made it unpredictable; it was "as likely to do the crazy as the sane thing at any given moment." But James went even further, noting that an injured brain, missing component parts that have been damaged or destroyed entirely, might be seen as "virtually a new machine." This new machine may initially perform abnormally, with psychopathological symptoms, but as neurological cases repeatedly demonstrated, this new brain often found its way back to normality. "A brain with part of it scooped out is virtually a new machine, and during the first days after the operation functions in a thoroughly abnormal manner. As a matter of fact, however, its performances become from day to day more normal, until at last a practiced eye may be needed to suspect anything wrong."[19] James's colleague Peirce had similarly linked pathology with novelty. When our brain is injured, Peirce noted, we act in "new ways." Usually this entails a state of mental illness, but Peirce was willing to admit that in special cases "disease of the brain may cause an improvement in the general intelligence."[20] The inherent plasticity of the brain revealed itself best in the injured brain, but the ability to reorganize and defend against damage was linked to the cognitive flexibility of human minds. Intelligence was, in a sense, considered to be a consequence of a certain disorganization and unpredictability, and potentially even pathological disorder might explain the leaps of a genius intelligence.

Clinical and experimental research on the brain in the early twentieth century was systematically exploring the ability of the brain to *reorganize* in the face of challenges—including the radical challenge of grave injury. The shock of disorder opened up the possibility of a new form of order that was not explicable in merely mechanical terms. As Constantin von Monakow put it in a book on brain localizations, the disruption of part of the brain led to a more general "shock" of the system (what he called *diaschisis*). "Any injury suffered by the brain substance will lead (just as lesions in any other organ) to a struggle [*Kampf*] for the preservation of the disrupted nervous function, and the central nervous system is always (though not always to the same degree) prepared for such a struggle."[21] The pathological turn awakened a total response, aimed not at a simple return to the original order but rather to an order that reestablished stable functioning in a new form. As von Monakow wrote (in a book cowritten with the French neurologist R. Mourgue), "It is a question of combat, of an active struggle for the creation of a new state of affairs, permitting a new adaption of the individual and its milieu. The whole organism intervenes in this struggle."[22]

Interwar neurological theory produced many theoretical models that emphasized the plastic nervous system that could adjust itself to constantly changing circumstances, and even to radical damage of the brain.[23] One influential figure in this field, the American psychologist Karl Lashley, sought to disprove localization theories of memory by systematically removing pieces of the brain of animal subjects and demonstrating the persistence of learned behaviors in mazes and other environments. Surprisingly the test animals were often still able to run the mazes efficiently, implying that the brain must have a way to restructure itself to compensate for the missing tissue. Lashley saw this capacity as continuous with the adaptive unity of the brain as a complex system, a system that integrated activity throughout its component parts: "The whole implication of the data is that the 'higher level' integrations are not dependent upon localized structural differentiations but are a function of some general, dynamic organization of the entire cerebral system." Lashley's term for this flexibility was *equipotentiality*, alluding to his theory that because the brain was a dynamic and ever-changing entity, it could *reorganize* itself when challenged with new obstacles or internal failures.[24] As he later wrote, "I have been struck by the fact that even very extensive destruction of brain tissue does not produce a disorganization. Behavior becomes simplified but remains adaptive." For Lashley intuitive, insightful, intellectual activity was dependent on this feature of the brain and nervous system—"its plastic and adaptive" nature.[25]

Perhaps the most important figure in this history of plasticity was Kurt Goldstein, who synthesized his own extensive clinical research on neurological defects in brain-injured patients with broader streams of thinking about organismic life itself. Using neurological concepts such as *diaschisis*, taken from von Monakow's work, as well as German holistic philosophy, especially Gestalt theory, Goldstein defined the organism less as a fixed organization or structure and more as a dynamic configuration constantly struggling to maintain coherent unity when challenged by the ever-changing conditions of life. Pathological states of being exhibited their own peculiar characteristics in neurological patients. When an organism's responses to the environment were "constant, correct, adequate," it could survive in its milieu. But in a condition of shock the organism was often led to "disordered, inconstant, inconsistent" actions, which created a "disturbing aftereffect."[26]

However, as Goldstein would argue, organisms have the capacity to modify themselves in the face of this disordered behavior, which he called a "catastrophe reaction." The organism's ability to reorganize in response to shock was, according to Goldstein, just one way that this organism sought unity in moments of disruption. The catastrophes that were labeled pathological were in fact continuous with the dynamic of a healthy, normal body as it struggled to maintain its equilibrium. That is, the organism was always seeking unity and order: "Normal as well as abnormal reactions ('symptoms') are only expressions of the organism's attempt to deal with certain demands of the environment."[27]

And so Goldstein writes that even the normal, healthy life of the organism can be considered a series of what he calls "slight catastrophes" (*leichter Katastrophenreaktionen*), where inadequacies are first confronted and then "new adjustments" or "new adequate milieu" are sought to respond to this lack. The more serious catastrophic breakdown differs only in the scale and intensity of the reaction. The whole organism is really a potential unity always falling into states of shock, and it must, over and over again, creatively establish new order to overcome these shocks: "If the organism is 'to be,' it always has to pass again from moments of catastrophe to states of ordered behavior."[28] The essential plasticity of the organism was of course most clearly visible in the brain because it was the center of organization and integration and therefore the site of reorganization and reconstitution.

However, the same structures of crisis and reorder can be tracked in interwar psychology in theories of creative and adaptive thinking. In Gestalt psychology most clearly, genuine intelligence was defined as the ability to solve difficult problems with new perspectives that reorganized the experience of the subject to produce new arrangements. The word used for this sudden realization was *insight*. In his famous ape studies conducted during World War I, Wolfgang Köhler aimed to uncover the primordial conditions of intelligence, using primates as a way into the essence of human intelligence before it was overlaid with language and cultural knowledge. Köhler showed how insight (*Einsicht*) was achieved when the mind freed itself from one interpretation of the situation to discover a new one that resolved the tension—in these cases, to see objects as tools that could lead to food that had been placed out of direct reach. For Köhler intelligence was revealed by the ability to "detour" away from the direct path in order to understand how to circumvent obstacles.[29] Psychological insight had its "isomorphic" counterpart in the organism itself: "Human beings construct machines and compel natural forces to operate in defined and controllable ways. A system of this sort is one of 'fixed' connections in the sense of analytical mechanics." But a *dynamic* system was capable of reorganization and adaptation. Köhler's example was the developing organism: "Many embryos can be temporarily deformed (artificially) in ways utterly incongruous with their racial history and nevertheless regain by a wholly other route the same developmental stage reached by another, untouched embryo."[30]

The investigation into "productive thinking" emphasized the novelty of analogical extensions of established knowledge into radically new zones of understanding.[31] Köhler's Gestalt colleague Kurt Koffka, for example, argued that the mind was not limited to combining and recombining the "reaction pathways" and memories that it already possessed. The mind had, he said, a fundamental "plasticity" that allowed for radical novelty and therefore real progress in thinking.[32] The Swiss experimental psychologist Edouard Claparède perfectly captured this relationship between automatisms of habitual thinking and radical liberty characteristic of intelligence; he also echoed Goldstein's idea of the catastrophic reaction: "The rupture of equilibrium, when reflexes or habits are not available to intervene, is not reestablished automatically, and we are momentarily disadapted. It is intelligence which takes on the task of readapting us."[33]

HERE, AT THE threshold of the computer age, it was understood that the brain's open structure precluded any easy reduction of human cognition to some forms of automatic "machinery." Yet the most advanced computing devices of the period were analog machines, such as Vannevar Bush's Differential Analyzer (ca. 1930) and the hybrid electromechanical Mark I at Harvard (1940–44). However complex these technological machines were, they clearly incarnated the principle of *automaticity*: constructed as a series of physically coupled moving elements, these analog devices were arranged to represent directly the relationships of various mathematical or other determined functions. This seemed to preclude any serious form of AI that could mimic intelligent life. As Vladimir Jankélévitch put it in 1931 in a book on Bergson, "With a perfect machine there is never any deception, but also never any surprise. None of those miracles which are, in a way, the signature of life."[34]

So when Turing first imagined the "universal" digital computer in a 1936 mathematical paper it was a strange interloper in the field of automatic computing precisely because it had *no intrinsic organization* of its own. As a simple machine with nothing more than the capacity to manipulate two symbols on a moving tape, this *digital* computer was defined as a radically open machine whose sole purpose was to take on the configurations of other discrete-state machines. What Turing invented was a machine that automatically mimicked (virtually, in coded representations) the successive states of *another* automatic machine. Soon enough, of course, the Turing machine became much more than that. The binary logic of the digital computer was quickly applied to synaptic connectivity in the brain. Turing's main point—that a digital computer's operation was governed by its *logical* and not physical organization—only strengthened the analogy of computer and brain in this period. For some the brain was in essence a digital computer instantiated by the sequential firing of neurons, which were analogous to the mechanical switching of relays or the processing of electrical impulses. The physical substrate of the Turing machine was irrelevant to its logical operation.[35]

Critics of the mechanistic worldview zeroed in on the rigid automaticity implied by the computer analogy. Machines, wrote

Georges Canguilhem in a 1948 essay, could only affirm rational "norms" of consistency and predictability. In contrast living beings were capable of *improvisation*; in its creative drive to survive, "life tolerates monstrosities." Pointing to the great plasticity of the nervous system, Canguilhem noted that if a child suffers a stroke that destroys an entire half of the brain, that child would not suffer aphasia (as is the case with many brain injuries later in life) because the brain reroutes language performance to other regions in order to compensate for the damage. Organisms are stable as *unities* precisely because their organization is *not* fixed into any one rigid structure; they are open, and thus equipped to surmount even a traumatic loss of functions in some cases. However, as Canguilhem declared, "There is no machine pathology."[36] And of course for Canguilhem the pathological state of illness revealed the living being's power to create new norms in crisis situations.[37] Michael Polanyi echoed this observation, writing, "The organs of the body are more complex and variable than the parts of a machine, and their functions are also less clearly specifiable." He noted that a machine knows only "rules of rightness" and can never accommodate error or "failures"—unlike living bodies that are capable of such radical transformation.[38] Or, as the systems theorist Ludwig von Bertalanffy put it, unlike a living, open system, a "mechanized organism would be incapable of regulation following disturbances."[39]

However, with the rise of cyberneticists came the belief in a wholly new kind of machine technology, a flexible, adaptive one that would mimic the vital improvisation of the organism. Yet the cybernetic machines (examples included self-guided missiles and other servomechanisms) were, it seems, still governed by the logic of automaticity, despite their ability to correct their behavior through negative feedback circuits. Adaptive responses to environmental changes were fully determined by the engineered system. The cybernetic being had specific embedded "goals" and a fixed organization of sensors, processors, and effectors to guide it to these ends. As a French neurologist put it in 1955, "The rigidity of electrical mechanisms deployed in machines or robots appeared from the start far removed from the variability, the plasticity of those in the nervous system."[40] Was there any possibility of genuine plasticity in the cybernetic machine? Could such a machine exhibit genuine pathologies in the sense that Goldstein and Canguilhem gave the term?[41]

The cybernetician W. Ross Ashby investigated the possibility of such a pathological machine, one that would then be capable of truly novel and unexpected behavior. In a notebook fragment from 1943 we find him reading William James. Ashby was particularly interested in passages from James where the rigid "machine" is compared to the organic and open structure of the nervous system. As we saw, for James this system paradoxically exhibited *both* fixity of structure and an open-ended, adaptive plasticity.[42] The notebooks also show that Ashby was reading Charles Sherrington's revolutionary work on neural integration, published in 1906.[43] Ashby zeroed in on passages that described the dynamic nature of the nervous system. Like James, Sherrington located the great advantage of the nervous system's fragility: by virtue of being so sensitive to disruption, the organism was highly sensitive to changes in its environmental milieu.[44] Inspired by these ideas, Ashby set out to construct a machine that could behave like an open system, behave, that is, like a determinate structure that was nonetheless capable of reorganization.

Ashby admitted in 1941 that artificial machines that could change their organizations were "rare."[45] He was searching for the machine's missing quality, that "incalculable element" (in the words of James) capable of producing novel actions. If machines were defined, as Polanyi had argued, by their calculability, Ashby's project would be impossible. But he eventually hit upon a new way of thinking about machines and plasticity. In essence he realized that for a determinate machine to be radically open to new forms of organization, it had to be capable of becoming, in a way, a wholly different machine. *Failure*, or breakdown, which was already inevitable in any concrete machine, turned out to be the key idea. For Ashby "a break is a change in organization."[46] A cybernetic machine, one that continually tried to find equilibrium, could at times find itself in a condition where maintaining homeostasis became impossible. However, if that homeostatic being was constructed so that it would, in these moments, break down in some fashion when equilibrium was not being achieved, then (theoretically at least) this machine could be said to acquire a new organization, and therefore new possible paths to equilibrium. As Ashby wrote, "After a break, the organization is changed, and therefore so are the equilibria. This gives the machine fresh chances of moving to a new equilibrium, or, if not, of breaking again."[47] Using the language of neurophysiology in this period, we could say that the internal failure of the machine was a form of shock to the system. Ashby pointed the way, then, to a new form of *cybernetic plasticity* that flowed from the basic weakness of any machine, namely the inevitability of failure.

Ashby's rethinking of the human brain followed from these ideas on the physical forms of adaptive behavior. The brain was not only dynamic and flexible in its organization; it was also a special kind of "machine" made up of individual elements (neurons) that were in essence unreliable, in that they would stop working at certain thresholds until they could recover and begin again to reconnect with other neurons. Due to their latency period, individual neurons were, so to speak, constantly appearing and *disappearing* from the system as a whole. The challenge for cybernetics was to model this form of productive failure within artificially constructed intelligent machines.[48]

The link between brain science and cybernetics offers a new way of contextualizing the origins of AI. Turing opened the field, conceptually at least, when he published "Computing Machinery and Intelligence" in 1950, introducing the famous "imitation game" to the public. Turing's project, as it is usually understood, was to think about how a digital computer could successfully simulate human intelligence by modeling thought in algorithmic processes. However, in 1946, in the midst of developing Britain's first stored program digital computer, the Automatic Computing Engine, Turing wrote to Ashby. In that letter Turing admitted that he was "more interested in the possibility of producing models of the action of the brain" than in any practical applications of the new digital computers inspired by his work.[49] His interest in models of the nervous system and brain in fact indicate a turn from the strict notion of machine automaticity introduced in his formal mathematical work on computability toward a new interest in the dynamic organization and reorganization of organic systems.

In 1936 Turing had envisioned the universal computer as, in theory, a perfectly automated, stand-alone machine. However, by 1948, after much experience with real computing during and after World War II, he was thinking more and more about the relationship between brains, human intelligence, and these new computers. Turing noted that by definition any machine can be "completely described" by listing all its possible configurations: "According to our conventions the 'machine' is completely described by the relation between its possible configurations at consecutive moments. It is an abstraction that cannot change in time." Yet he imagined that if something *interfered* with this machine, the machine would in essence be modified; it would become a new machine. Turing then suggestively noted that human learners might well be considered "machines" that have been constantly modified by massive interference, namely through teaching and communication, and sensory forms of stimuli.[50]

Turing remarked, "It would be quite unfair to expect a machine straight from the factory to compete on equal terms with a university graduate. The graduate has had contact with human beings for twenty years or more. This contact has been modifying his behaviour throughout that period. His teachers have been intentionally trying to modify it." The "routines" of thought and action in a student have been "superimposed on the original pattern of his brain." Now it was not the case that the human became an automaton in this process of "programming." Rather the open machine that is subject to such modification by interference is the one that becomes capable of real *creativity*. With new routines the human-machine can, Turing said, "try out new combinations of these routines, to make slight variations on them, and to apply them in new ways."[51] And crucial to genuine intelligence was the freedom from automatic routine: "If a machine is expected to be infallible, it cannot also be intelligent. The machine must be allowed to have contact with human beings in order that it may adapt itself to their standards." Intelligence, Turing wrote, "consists in a departure from the completely disciplined behavior involved in computation."[52] Therefore the intelligence of a computer (or a human) was not measured by its computational prowess but by its radical openness to interference—from outside, most obviously, but also, to a certain extent, from within.

Turing's hypothesis was that the infant human brain should be considered an *unorganized* machine that acquires organization through suitable "interference training": "All of this suggests that the cortex of the infant is an unorganized machine, which can be organized by suitable interference training. . . . A human cortex would be virtually useless if no attempt was made to organize it. Thus if a wolf by a mutation acquired a human cortex there is little reason to believe that he would have any selective advantage."[53] An isolated, solitary human brain would make no progress because it needs a social milieu of other human beings in order to learn and create. Turing's project for AI was not, as we usually think, a project to dissect the operations of cognition and translate them into programmable routines. Instead the goal was to model a mostly open, pliant brain that *transforms itself* as nature and culture impress themselves upon it. "Our hope," he wrote, "is that there is so little mechanism in the child-brain that something like it can be easily programmed."[54] Human intelligence, like computer intelligence, is in its essence already *artificial*, imprinted, that is, from outside onto the plastic architecture of the brain/computer.

Turing's American counterpart, John von Neumann, was equally fascinated by the robust and flexible behavior of organic nervous systems and brains. Appointed to oversee the development of an electronic computer at the Institute for Advanced Study at Princeton from 1945 to 1951, von Neumann looked to neurophysiology for inspiration. That the brain and the nervous system exhibited an amazing robustness was, he observed, in stark contrast to the immense fragility of the new high-speed computers then being constructed out of undependable mechanical relays, telephone switches, or vacuum tubes. In his early reflections on the computer, von Neumann was careful to draw attention to the critical *differences* between digital machines and the nervous system.[55] Yet he himself was drawn to the nervous system as a way of thinking of the computer as something more than a sequential calculating device. One of the most important marks of the natural communication and control system was, he saw, its inherent flexibility and hence dependability: "It is never very simple to locate anything in the brain, because the brain has an enormous ability to re-organize. Even when you have localized a function in a particular part of it, if you remove that part, you may discover that the brain has reorganized itself, reassigned its responsibilities, and the function is again being performed."[56]

In 1948 von Neumann gave his celebrated lecture "The General and Logical Theory of Automata" at the Hixon Symposium held at Cal Tech. This interdisciplinary gathering on the topic of the brain is one of the key founding moments of cybernetics. Some of the most influential researchers in psychology were in attendance there, including holistic thinkers such as Köhler and Lashley. In fact von Neumann seemed sensitive to this tradition in his own presentation. He was interested in developing artificial computing machines that mimicked the robustness of natural, living systems, and he pointed to just the kind of nervous system behavior studied by the neurophysiologists. As he explained, the organism was always challenged by unpredictable but inevitable *errors*. How did the nervous system in particular maintain itself despite the pressure of pathological circumstances? In the discussion of von Neumann's paper it was noted that we must "not only be able to account for the normal operation of the nervous system but also for its relative stability under all kinds of abnormal situations."[57]

Error was linked, in von Neumann's work, to plasticity. If a complex computer was going to be flexible and stable across error and failure it was important that "error be viewed . . . not as an extraneous and misdirected or misdirecting accident, but as an essential part of the process under consideration."[58] Somehow the computer would need to have the ability to "automatically" sense where errors were occurring and simultaneously reorganize the system as a whole to accommodate those errors. It seems clear that von Neumann was alluding directly to key concepts taken from theoretical biology, what Lashley called "equipotentiality" and Bertalanffy understood by "equifinality." Von Neumann looked to how natural organisms could *reorganize* themselves when they faced what he labeled "emergency" conditions.[59]

FROM ITS ORIGIN the digital computer was a machine that was always awaiting its determination. To put it another way, the computer needed to contain what the postwar philosopher of technology Gilbert Simondon called "a margin of indetermination." For Simondon the perfect machine could never be perfectly automatic. If it was going to be open to information and therefore capable of genuine communication, then the machine, like the organism, must be in some measure plastic: "The true perfection of machines . . . does not correspond to an increase in automation, but on the contrary to the fact that the functioning of a machine harbors a certain margin of indetermination." If, as Descartes understood centuries ago, the human being is an automaton of sorts, determined by its education, training, language, and culture, it is nonetheless a special kind of automaton. As Simondon put it, "The human being is a rather dangerous automaton, who is always risking invention and giving itself new structures."[60] The challenge of creating any intelligent automata is the challenge of modeling indetermination, disruption, failure, and error. In an age of automation that figured cognition itself as an automatic procedure, it was doubly important to show that the automaticity of the thinking being was predicated on a more foundational *plasticity*. Lewis Mumford explained this in 1964: "Let me challenge the notion that automation is in any sense a final good. If the human organism had developed on that principle, the reflexes and the autonomic nervous system would have absorbed all the functions of the brain, and man would have been left without a thought in his head. Organic systems [have] . . . the margin of choice, the freedom to commit and correct errors, to explore unfrequented paths, to incorporate unpredictable accidents . . . to anticipate the unexpected, to plan the impossible."[61]

In our own age of automation, where the automaticity incarnated by digital technologies structures the conceptual foundations of cognitive science, it is once again time to rearticulate human nature in terms of what is not automatic. Our digital brains—brains modeled on and simulated by computers and increasingly *formed* by repeated interactions with our digital prostheses—will reveal their genuine plasticity only when they rediscover the power of interrupting their own automaticity.

# Endnotes

[1] The opening sections of this essay draw on my "Penser l'automaticité sur le seuil du numérique," in *Digital Studies: Organologie des savoirs et technologies de la connaissance*, ed. Bernard Stiegler (Paris: Editions FYP, 2014).

[2] Morana Alač, *Handling Digital Brains: A Laboratory Study of Multimodal Semiotic Interaction in the Age of Computers* (Cambridge, MA: MIT Press, 2011).

[3] Thomas P. Trappenberg, *Fundamentals of Computational Neuroscience* (Oxford: Oxford University Press, 2010).

[4] On the intertwined history of cognitive science and artificial intelligence, see Paul Edwards, *The Closed World: Computers and the Politics of Discourse in Cold War America* (Cambridge, MA: MIT Press, 1997); Jean-Pierre Dupuy, *On the Origins of Cognitive Science: The Mechanization of Mind*, trans. M. B. DeBevoise (Cambridge, MA: MIT Press, 2009).

[5] Human Brain Project, "Overview," https://www.humanbrainproject.eu/discover/the-project/overview (accessed February 12, 2015).

[6] U.S. Department of Health and Human Services, National Institutes of Health, "Brain Research through Advancing Innovative Neurotechnologies (BRAIN)," braininitiative.nih.gov (accessed February 12, 2015).

[7] Catherine Malabou, *The Ontology of the Accident: An Essay on Destructive Plasticity*, trans. Caroline Shread (London: Polity, 2012).

[8] Rajagopal Ananthanarayanan, Steven K. Esser, Horst D. Simon, and Dharmendra S. Modha, "The Cat Is Out of the Bag: Cortical Simulations with $10^9$ Neurons, $10^{13}$ Synapses," *Proceedings of the Conference on High Performance Computing Networking, Storage and Analysis*, article 63, http://ieeexplore.ieee.org/stamp/stamp.jsp?tp=&arnumber=6375547 (accessed June 2, 2015).

[9] Paul Merolla et al., "A Digital Neurosynaptic Core Using Embedded Crossbar Memory with 45pJ per Spike in 45nm," *IEEE Custom Integrated Circuits Conference*, September 2011. http://ieeexplore.ieee.org/stamp/stamp.jsp?tp=&arnumber=6055294 (accessed June 2, 2015).

[10] Mrigank Sharad et al., "Proposal for Neuromorphic Hardware Using Spin Devices," Cornell University Library, July 18, 2012, http://arxiv.org/abs/1206.3227v4 (accessed June 2, 2015).

[11] Dharmendra S. Modha et al., "Cognitive Computing," *Communications of the ACM* 54 (2011): 65.

[12] B. G. Farley and W. A. Clark, "Simulation of Self-Organizing Systems by Digital Computer," *Transaction of the IRE* (1954): 7.

[13] René Descartes, *Treatise on Man*, in *The Philosophical Writings of Descartes*, trans. John Cottingham, Robert Stoothoff, and Dugald Murdoch, 3 vols. (Cambridge: Cambridge University Press, 1985), 1: 104.

14 René Descartes, *Discourse on Method*, in *Philosophical Writings*, 1: 119.

15 David Bates, "Cartesian Robotics," *Representations* 124 (Fall 2013): 43–68.

16 C. S. Peirce, *Collected Papers of Charles Sanders Peirce*, 8 vols. (Cambridge, MA: Harvard University Press, 1931–58), 2: para. 59.

17 C. S. Peirce, letter to A. Marquand, December 30, 1886, in C. Kloesel et al., eds., *Writings of Charles S. Peirce: A Chronological Edition* (Bloomington: Indiana University Press, 1993), 5: 421–22. An image of the original letter with the circuit design diagram is on 423.

18 William James, *Principles of Psychology*, 2 vols. (New York: Henry Holt, 1890), 1: 105.

19 Ibid., 1: 139–40, 143.

20 C. S. Peirce, *Contributions to the Nation*, 3 vols. (Lubbock: Texas Tech University Press, 1975–87), 1: 144.

21 Constantin von Monakow, *Die Lokalisation im Grosshirn und der Abbau der Funktion durch kortikale Herde* (Wiesbaden: J. F. Bergmann, 1914), 30.

22 Constantin von Monakow and R. Mourgue, *Introduction biologique à l'étude de la neurologie et disintegration de la function* (Paris: Félix Alcan, 1928), 29.

23 The following discussion of neurophysiology takes up the argument made in my "Unity, Plasticity, Catastrophe: Order and Pathology in the Cybernetic Era," in *Catastrophes: History and Theory of an Operative Category*, ed. Andreas Killen and Nitzen Lebovic (Berlin: De Gruyter, 2014), 32–54.

24 Karl Lashley, *Brain Mechanisms and Intelligence: A Quantitative Study of Injuries to the Brain* (Chicago: University of Chicago Press, 1929), 176.

25 Karl Lashley, "Persistent Problems in the Evolution of Mind," *Quarterly Review of Biology* 24 (1949): 33, 31. The Soviet psychologist and neurologist Alexander Luria developed a similar theorization of neural plasticity. In his studies of patients who had suffered various brain injuries, Luria observed the loss of certain cognitive abilities, but he also observed repeatedly that the brain had the startling ability to reorganize itself in order to compensate for the loss of functions after stroke or accident. Luria noted that numerous studies showed "the high degree of plasticity shown by damaged functional systems, due to dynamic reorganization and adaptation to new circumstances and not to regeneration and restoration of their morphological integrity." A. R. Luria, *Restoration of Function after Brain Injury*, trans. Basil Aigh (New York: Macmillan, 1963), 33. See as well Laura Salisbury and Hannah Proctor's essay in this volume.

26 Kurt Goldstein, *The Organism: A Holistic Approach to Biology Derived from Pathological Data in Man* (1934; New York: Zone, 1995), 48–49. An excellent account of Goldstein's work and the German contexts of theoretical biology can be found in Anne Harrington, *Reenchanted Science: Holism in German Culture from Wilhelm II to Hitler* (Princeton, NJ: Princeton University Press, 1996).

27 Goldstein, *The Organism*, 52, 35.

28 Ibid., 227, 388.

29 Wolfgang Köhler, *Intelligenzprüfungen an Menschenaffen* (Berlin: Springer, 1921).

30 Wolfgang Köhler, "Some Gestalt Problems," in *A Sourcebook of Gestalt Psychology*, ed. Willis D. Ellis (London: Kegan Paul, 1938), 55, 66.

31 Karl Duncker, *Zur Psychologie des produktiven Denkens* (Berlin: Springer, 1935); Max Wertheimer, *Productive Thinking* (Ann Arbor: University of Michigan Press, 1945).

32 Kurt Koffka, *The Growth of the Mind: An Introduction to Child Psychology* (New York: Harcourt Brace, 1925), 125.

33 Edouard Claparède, "La genèse de l'hyopthèse: Étude expérimentale," *Archives de Psychologie* 24 (1934): 5.

34 Vladimir Jankélévitch, *Henri Bergson* (Paris: Presses Universitaires de France, 1989)

35 Warren McCulloch and Walter Pitts, "A Logical Calculus of the Ideas Immanent in Nervous Activity," *Bulletin of Mathematical Biophysics* 5 (1943): 115–33; Norbert Wiener, *Cybernetics* (Cambridge, MA: MIT Press, 1948); Warren McCulloch and John Pfeiffer, "Of Digital Computers Called Brains," *Scientific Monthly* 69 (1949): 368–76.

36 Georges Canguilhem, "Machine et organisme" (1948), in *La connaissance de la vie*, 2nd ed. (Paris: Vrin, 1980), 118.

37 Georges Canguilhem, *On the Normal and the Pathological*, trans. Carolyn R. Fawcett (New York: Zone, 1991).

38 Michael Polanyi, *Personal Knowledge: Towards a Post-Critical Philosophy* (Chicago: University of Chicago Press, 1958), 359, 328.

39 Ludwig von Bertalanffy, *Das Biologische Weltbild* (Berne: A. Franke, 1949), 29.

40 Théophile Alajouanine, *L'homme n'est-il qu'un robot? Considérations sur l'importance qu'a l'automaticité dans les functions nerveuses. Discours inaugural prononcé à la séance du Congrès des psychiatres et neurologistes de langue française, à Nice, le 5 septembre 1955* (Cahors: A. Coueslant, 1955), 11.

41 We can note that Bertalanffy had criticized cybernetics precisely for its misguided use of closed systems to model the fundamentally open structures of natural organisms. Ludwig von Bertalanffy, *General System Theory: Foundations, Development, Applications* (New York: George Braziller, 1968).

42 W. Ross Ashby, *Journal*, 1523–24, W. Ross Ashby Digital Archive, ross-sashby.info/index.html (accessed July 24, 2013).

43 Charles S. Sherrington, *The Integrative Action of the Nervous System* (New Haven, CT: Yale University Press, 1906).

44  Ashby, *Journal*, 1906–9.

45  Ibid., 1054.

46  W. Ross Ashby, "The Nervous System as Physical Machine: With Special Reference to the Origin of Adaptive Behaviour," *Mind* 56 (1947): 50. On Ashby's theoretical work and engineering projects in homeostasis, see Andrew Pickering, *The Cybernetic Brain: Sketches of Another Future* (Chicago: University of Chicago Press, 2009), especially chapter 4

47  Ashby, "The Nervous System as Physical Machine," 55.

48  Ibid., 57–58.

49  Alan Turing, letter to W. Ross Ashby, ca. November 19, 1946, quoted in Jack Copeland, introduction to Alan Turing, "Lecture on the Automatic Computing Engine (1947)," in Alan Turing, *The Essential Turing: Seminal Writings in Computing, Logic, Philosophy, Artificial Intelligence, and Artificial Life plus the Secrets of Enigma*, ed. B. Jack Copeland (Oxford: Oxford University Press, 2004), 374.

50  Alan Turing, "Intelligent Machinery" (1948), in Turing, *The Essential Turing*, 419.

51  Ibid., 421.

52  Alan Turing, "Computing Machinery and Intelligence," *Mind* 59 (1950): 459.

53  Turing, "Intelligent Machinery," 424.

54  Turing, "Computing Machinery," 456.

55  John von Neumann, *The Computer and the Brain* (New Haven, CT: Yale University Press, 1958).

56  John von Neumann, *Theory of Self-Reproducing Automata*, ed. Arthur W. Burks (Urbana: University of Illinois Press, 1966), 49.

57  Ibid., 323.

58  Ibid., 329. I have discussed von Neumann's theory of error in "Creating Insight: Gestalt Theory and the Early Computer," in *Genesis Redux: Essays in the History and Philosophy of Artificial Life*, ed. Jessica Riskin (Chicago: University of Chicago Press, 2006), 237–59. See as well Giora Hon, "Living Extremely Flat: The Life of an Automaton. John von Neuman's Conception of Error of (In)animate Systems," in *Going Amiss in Experimental Research*, ed. Giora Hon, Jutta Schickore, and Friedrich Steinle, Boston Studies in the Philosophy of Science, no. 267 (N.p.: Springer, 2009), 55–71.

59  Von Neumann, *Theory of Self-Reproducing Automata*, 71, 73.

60  Gilbert Simondon, *Du mode d'existence des objets techniques*, rev. ed. (Paris: Aubier, 1989), 11–12.

61  Lewis Mumford, "The Automation of Knowledge: Are We Becoming Robots?," *Audio-Visual Communication Review* 12 (1964): 275.

*Joseph Dumit*

# 8 Plastic Diagrams: Circuits in the Brain and How They Got There

## Poring Over Brain Diagrams

TODAY WE TALK easily about the similarities between brains and computers, between programs and thought processes, between neurons and neural networks.[1] We speak of recovering addicts "reprogramming" their minds. We talk about "hard-wired" fears and urges. We order additional "memory" for our laptops.

The computer—the mechanistically functioning contraption—has become a go-to metaphor for the brain, whose complexity nonetheless eludes the orderly, mechanical rules one would expect of such a machine. There's the sense that we can easily understand the computer, or at least its programs, because it is mechanical: it follows rules; it is rational. But what about the brain? Does it "appear alive" merely because its computer-like machinery is so complex that we lose track of what's going on and throw our hands in the air? That may be the view supported by the brain-computer metaphor—that inside our skulls there's nothing more than what amounts to a mere mess of electrical circuits but that they're so intricate, so hard to tease apart that we've decided to think of them as "mysterious," simulating life. I've begun to suspect that we celebrate the brain as plastic in order for it to be something more than a computer.

COGNITIVE SCIENTISTS, PERHAPS more often than the rest of us, like to think of brains as computers. While conducting ethnographic fieldwork on brain imaging, looking at how experiments were designed, subjects selected, images produced and circulated, I was struck by how often researchers suggested there are circuits in the brain. And as I watched them return to this analogy over and over again, I noticed a third element that kept appearing, lingering alongside everything, utterly taken for granted: flowcharts. Drawings of the circuits that purportedly make up our brain. Textbooks and journal articles often juxtaposed these box-and-arrow diagrams of our brain circuitry with brain images, as if the diagrams were explanations of the evidence.

The PET scan researchers I spent time studying in the 1990s were continually trying to identify these circuits in the brain in order to connect them to mind. One summer, for instance, I followed around psychiatrist K.T. and his colleagues at a university, spending hours with them in the computer room where they processed brain images, in the imaging center where subjects were given radioactive tracers, performed tasks, and then had their brain scanned, and in meeting rooms where they discussed their new data. During these observations I found myself in a room with K.T. and four collaborators from pharmacology, speech therapy, neuroanatomy, and neuroscience as they prepared to write a grant proposal. They were poring over preliminary brain scans scattered across the table, attempting to piece together a hypothesis about a reading disorder. As they peered at the colorful brain images before them, they pointed out possible hotspots that suggested brain activity, consulted journals online to figure out what brain regions these hotspots might correspond to, and then proceeded to sketch the boxes and arrows on paper to propose circuits and searched the literature for possible theories on reading disorders that could be described by each hypothesized circuit. These interdisciplinary researchers were communicating with each other by drawing and editing each other's flowcharts. The diagrams as they evolved were both anatomical maps and functional circuits. In other words, every single box in the scribbled flowcharts was meant to correspond to a particular area of the brain, but also to represent the execution of a particular "mental step." These prospective grantees were actively evolving flow diagrams that would serve to explain their current data and present a model for future experiments. This practice was thus creative and constraining at the same time. Yet the question was never whether brains had circuits, but only which circuits were involved.

When I asked how they came to use box-and-arrow diagrams to represent their speculations, they seemed perplexed. They had always done so. It just made sense. And it worked. When I tried to study the interaction of visual metaphors in designing the imaging algorithms, psychological task protocols, and brain circuitry theories, the answer I constantly received from the researchers was "We used what was available." Inquiring into the history of the software programs that turned stochastic data emitted from brains into colorful pictures that are often interpreted as evidence of circuits in our minds, I was stumped by the circularity of it all. The researchers I studied used software programs that assumed that the noisy, sometimes diffuse areas of activity were produced by small, specific regions of the brain (represented by a flowchart's boxes), separated by gulfs of space (presumably traversed by a flowchart's arrows). It turned out that the neuroimaging results that so nicely seemed to demonstrate circuits were made by algorithms with built-in assumptions that circuits were present.[2] Here the software and the neurocognitive researchers agreed implicitly.

The circuits and their diagrams, in other words, were flowcharts that invoked computer chips or computer programs, and this relation was one of final explanation, because once we had the complete circuit diagrammed, we would fully understand. Assuming that it is possible to compare brains to circuits is a way to make the brains seem more functional, manageable, and understandable. No one is claiming that we are anywhere close to a complete diagram of course, but the implication always was that if it were possible one day to model the brain-mind as a computer, then we will have reached our goal—because computers are not really mysterious, only sometimes very complex, and because inside computers are only small circuits, mechanisms, that are not mysterious at all but rather the most commonsense examples of logic and rationality, the foundations of our agency.

## Rethinking Flowcharts

MY ETHNOGRAPHY OF contemporary science thus led to historical questions: Where and when did the notion of the brain as computer arise? From whom did programmers who created the software used to interpret data from neuroimaging devices borrow this notion of circuits? How did the drawing conventions themselves evolve? To what extent is the brain = computer equation today sustained by the fact that the same diagrammatic practices and convention are used for each? In other words, was there a period when the conventions and assumptions underlying circuits and flowcharts seeped, unnoticed, into the way we think about and study brains? In this chapter I want to begin a genealogy of these diagrams, inquiring into where these practices of drawing circuits that melded brain, mind, and computer came from, and how they changed along the way.

At the origin of contemporary flowcharts are two creative diagramming practices from the 1940s that emerged out of attempts to model the behavior of digital circuits in time. These diagrams grew out of problems and concerns with the timing and temporality and pathology of circuits. I've been surprised over and over at how strange these researchers found running computers, running digital circuits. They were uncanny, weird, and irrational. Yet they were clearly logical. One could follow them along as they followed their instructions, step by step. And yet the result was often not rational. They would get caught in a loop and have no way out, akin to people getting stuck in a rut or becoming obsessive. They would ignore everything else except their programming, like subjects of ideology. Little errors would compound into big ones. They had no way of reflecting on whether they were running the right program. In other words, different running computers were great analogs for types of personalities. They were logical each in their own way, "patho-logical." Circuits (in our brains or minds) were seen as part of what we as humans have to deal with and learn to live in relation to. It was only later that computers and whole minds/brains came to be directly equated.

Reading these early researchers I've been struck by how easy it is to be anachronistic and think that they talk about computers and draw circuits the way we do. But they don't. The first thing to ponder is how they could draw a little logical circuit, a whole simple program, and then find it mysterious. In my curious reading, which follows these moments of mystery only, they were fascinated by the consequences of following the circuit along its path, *in time*, step-by-step, in order to see where it led them: whether to a solution, or a loop, or something quite unexpected (especially when feedback between two or more circuits was involved).[3]

The main idea of this chapter is to recover how for decades some people found computers wily because they were logical circuits, and at issue was what it meant to be in this particular circuit or that one, not whether a brain or mind was actually a circuit or a computer in general. To juxtapose a circuit and a neurosis, or a mechanism and an affect, or a diagram and a pathology was not to reduce one to the other but to expand our understanding of each by recognizing how each pair might share a particular strange temporality. Neither human memory nor calculator memory could be taken for granted. Today many cognitive neuroscientists imply that reduction to mechanism or computer program or circuit is success; this is because most people think they completely understand how calculators or computers work. But in the 1940s–1960s computers and diagrams, circular systems and psychotic people, traffic jams and fascism equally boggled the mind despite being able to simulate them. There was no explaining away, only the modeling of problems that still remained problems.

## McCulloch's Existential Devices

*The McCulloch laboratory, more than any other, cared about the relations between cellular events and the processes of mind. The spirit of the laboratory was well characterized by the sign on the door. It read "Experimental Epistemology."*
—Robert Gesteland, "The Olfactory Adventure," in *Collected Works of Warren S. McCulloch*, ed. Warren S. McCulloch, Rook McCulloch and Heinz von Foerster, vol. 3 (Salinas, CA: Intersystems, 1989)

THE FIRST DIAGRAMMING practice I want to focus on is the McCulloch-Pitts logical neuron diagrams. Warren McCulloch trained in philosophy, psychology, neurophysiology, engineering, and psychiatry. He worked and thought with Norbert Wiener and chaired the wide-ranging Macy conferences on cybernetics (1946–53). As an engineer he thought constantly about the problems of real machines, and as a psychiatrist he thought constantly about problems that real humans have with their thinking—he found both uncanny. He was fond of discovering what he called "existential devices" in which an existing mechanism could be used to posit a possible material basis for a psychic structure.[4]

McCulloch was the coauthor with Walter Pitts of the original work defining the field of neural nets. Their 1943 paper was titled "A Logical Calculus of the Ideas Immanent in Nervous Activity," and it in fact took the nervousness of the nets quite literally.[5] The paper begins by taking up Rudolf Carnap's project on a "propositional calculus,"[6] literally a calculus of propositions about the world. Carnap wanted to logically work out relations among ideas in the world. What Pitts (who was Carnap's protégé) did was add time to Carnap's logical notation system so that they could create a "formal equivalence" to neural activity in time, creating what they called *temporal propositions*.[7] Starting with the idea that "the response of any neuron [is] factually equivalent to a proposition which proposed its adequate stimulus,"[8] they then accounted for the stimulus as proposed to have taken place at time T-1. The innovation in notation was that neurons acted in discrete time states (synaptic delays). If a neuron at time T had just received two or more shots, it will shoot out a signal to the neurons it is pointing to at time T+1. Each neuron fired on by two at that time will fire the next time, and so on (fig. 8.1). According to Michael Arbib's history of the paper, "Taking account of the delay in each neuron's response to its inputs—and thus resolving McCulloch's concerns about the possible temporal paradoxes in neural networks containing loops—was a key ingredient in the eventual development of a logical calculus of neural activity."[9]

For McCulloch each neuron that fires offers a proposition to the next neurons, a proposition that implies the neuron that fired on them.[10] This paper first proved that even these far too simple "neurons" could calculate anything that a Turing machine could (that is, they were computing machines). The first finding of the paper was that, to the extent that human neurons at least include the possibility of functioning like these ideal neurons (which they do), they can calculate all calculable things. This is the first existential fact about our neurons.

The conclusion to the paper is more startling, and little discussed. To understand such a neuronal circuit one cannot just look at it—this is crucial: one must enact it. These diagrams are nonsensical without a discrete, lived notion of time. Here McCulloch was concerned that if even simplistic "neurons" were to be

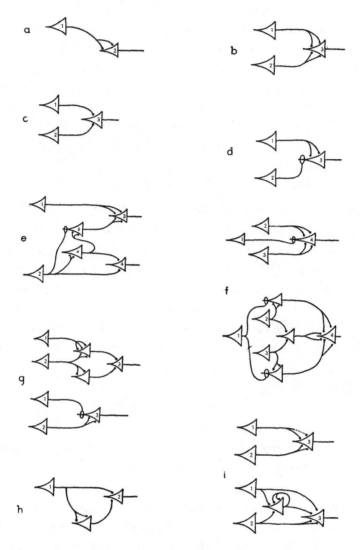

**Figure 8.1** McCulloch-Pitts "neurons." Each neuron fires from its flat side (to the right), and each neuron that is fired upon by two or more sources (bulb end of lines) will fire in the next time step. So in (a), if 1 fires, 2 will fire in the next step because two bulbs come from 1. In the case of (i) in the bottom right corner, the middle unnumbered neuron will fire on itself each time step 1 and 2 fire simultaneously. Thus it serves as a "memory" that 1 and 2 happened together at least once sometime in the past.
Source: McCulloch and Pitts, "A Logical Calculus," figure 1, p. 130.

referred to humans, then the question of the "subject" of these neurons needed to be posed. For instance, one of the key properties of a nervous net is that, looked at from the outside, they are deterministic forward in time but undetermined backward. That is, given the state of a net at time T, the state of the net at time T+1 is predictable. But the state of the net at time T–1 is not. In the case where two different neurons (A or B) could have caused the action, the previous state might have been either one. A third possibility is that there was a misfiring.

These nets, McCulloch said, were more than metaphors. Tara Abraham shows that these neural diagrams did not represent neurons; they presented "theoretically conceivable nets."[11] But, equally, this did not mean that the process itself was fully understood: "Man-made machines are not brains, but brains are a very ill-understood variety of computing machines."[12] Specifically McCulloch insisted that the point of view of the machine mattered and that we should consider ourselves as if we were a type of machine that functioned like a neural net. McCulloch was thus interested in what happens from the inside: if you were in such a circuit, enacting it, as you must, then you could find no difference among these three possibilities.

You could not know if the definite ringing in your ears is truly happening outside or not. You could not know if you really saw what you remember. Here was a logical basis for understanding tinnitus, hallucinations, delusions, and confusions. McCulloch is making an existential point: to the extent that our brains have structures like these nets (which they do), they can be structurally, logically delusional, hallucinatory, confused, and so on. And to the extent that mechanisms have subjects, they too must be thought through as neurotic, deluded, haunted, and so on.[13]

McCulloch termed this process of investigation "experimental epistemology," and it involved asking about the existential being of mechanisms and the mechanistic being of subjects. Experimental epistemology involved understanding each model as lively in its own manner, as a different species of thought and knowing, posing the question of different forms of life and knowing to different circuits.[14]

McCulloch and Pitts described, for instance, a memory neuron like the one in figure 8.2, which, once activated, will fire in every subsequent time state because at each moment it activates itself again. This circular action thus "represents" a memory not of the time it was activated but only of having been activated at some indeterminate time in the past: "Thus, our knowledge of the world, including ourselves, is incomplete as to space and indefinite as to time. This ignorance, implicit in all our brains, is the counterpart of the abstraction which renders our knowledge useful."[15]

A particular form of epistemology (here the abstraction of "the past in general") is matched to a particular form of neural net ("a single neural circle") as its existential device. As existential devices, suggested McCulloch, each circuit might be a model for a psychic structure, everyday ones as well as neuroses and psychoses. Different circuits suggest different therapies. In the case of the prospective but not retrospective deduction, he takes aim at some psychotherapies: "The psychiatrist may take comfort from the obvious conclusion concerning causality—that, for prognosis, history is never necessary."[16]

In another paper McCulloch explored a topology of circular preference.[17] If you are in it, and at this moment you prefer A, then the next moment you will prefer B (over both A and C), and then once you have B, you will prefer C (over A and B), and if you are in C, then you prefer A (over B and C), and so on in an endless loop. Such neural topologies are easy to build with circuit materials, and they have been identified in brains. McCulloch then wondered whether such a circuit, should it be at the heart of our motivational structure, would put us on an endless treadmill, preferring love to leisure, leisure to work, and work to love, in a never-ending yet always motivated cycle. This little tiny logi-

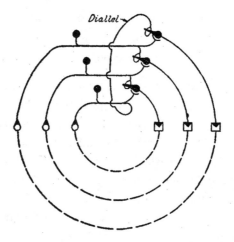

**Figure 8.2**  McCulloch's "irreducible nervous net." In this diagram the solid lines represent connections inside a body, and the dotted lines represent the outside environment. Due to the special arrangement of inhibitions (the loops around the bulb-like neurons), it forms a circular preference in which the top is preferred to the middle, the middle to the bottom, and the bottom to the top.
Source: McCulloch, "A Heterarchy of Values," figure 4, p. 92.

cal circuit, completely easy to see and draw and build and run, might hold a tragic insight into the human condition. If we were in such a circuit, our fate would be such a repetition, and herein he sought to understand our psychic life—full of headaches, he said.[18] To understand such a neuronal circuit one cannot just look at it—this is crucial: one must be in it running, enact it in time, trace it. This is similar to Norbert Wiener's claim that the brain "is not the complete analogue of the computing machine, but rather the analogue of a single run on such a machine."[19] These diagrams are nonsensical without a discrete, lived notion of time.

Time matters because McCulloch is sensitive to the psychic life of circuits, one that he emphasized many times: "Because light falling on a rod [in the eye] may or may not start a signal, that signal implies—but only implies—the light. . . . Thus what goes on in our brains implies—but only implies—the world given in sensation."[20] His point in this implication is that trust and doubt, psyche in general, is constitutive in life *because it runs in circuits.* Truth for us is more than physics. A signal is *both* physical (on/off) *and also* semiotic (true/false). He writes:

> Perhaps this calculus [of atomic propositions] does something more important, for it separates physics, for which the signals are only something that happens or else does not happen, from communication engineering, for which these same signals are also either true or else false. If you press on your eye you will see a light when there is no light. The signal is just as physical as ever, but because it arose in the wrong way or in the wrong place, it is a false signal, just as false as a ring on the telephone when lightning strikes the wires. It is because communication engineering deals with signals, true or else false, that neurophysiology is part of engineering, not merely of physics.[21]

We should pause here and note the importance of this claim, one missed by most commentators on McCulloch, who read him as a physicist, "on = true," "off = false," when in fact he is doing his best to distinguish himself from one.[22] He writes, "[The physical happening of a signal] is half the story! It is the essence of a signal that it proposes something. A signal is true if what it says happened did happen, otherwise it is false—for example, paresthesia or hallucination. Thus a signal has the mental—the logical properties of a proposition. By definition, a signal is a proposition embodied in a physical process."[23] It is best to keep in mind that there are no signals in general, only specific signals in specific places and times, signals that can lie.[24]

In 1949 McCulloch delivered a paper on "psychoneuroses" to psychiatrists, explaining to them how engineering systems with certain kinds of loops have what engineers call "gremlins," similar to how a problem in an automobile's carburetor can cause a problem in the engine block that causes a problem in the exhaust. Learning this "etiology" doesn't help figure out what to do about the exhaust.

One of his examples was causalgia, a condition in which snakebite on a finger causes tremendous pain first in the fingertip, then the hand, then arm, then spine, then brain, where it persists interminably. He showed a circuit arranged such that causalgia is not a misfiring but rather a form of action in the circuit that is created by the snakebite but sustains itself as a spiral action, best explained as a self-sustaining demon that doesn't have physical existence except as this temporal repetition, a soul that starts at one point of a large circuit and travels through it. Each temporal circuit, separated from the rest by gaps, thus functioned as a kind of "center of indetermination,"[25] with a subject, a form, a will, and a pathology all its own.

The year before, McCulloch gave a lecture to engineers, "The Brain as a Computing Machine," about the need to consider certain kinds of electrical systems as suffering from neuroses and benefiting from consideration of psychology. He thus posed to each discipline the necessity to attend to the form of posing problems of time, location, and liveliness that arise in the other discipline:

> In the neurotic brain you may find no general chemical reaction gone astray, nor any damaged cells, for when activity ceases, regeneration ceases. The most you might expect to find are some changed thresholds or connections—those little invisible differences which each of us acquires by use—the basis of our characters. The more we build negative feedback into machines, the more surely they will have neuroses. These diseases are demons with ideas and purposes of their own. Physicists have been known to curse them but they cannot be exorcised. If, instead of our variety of psychodynamic nonsense, you wish to think sensibly of them I would suggest, in all seriousness, that you start now to prepare a dimensional analysis of gremlins.[26]

In a future paper I will work through in detail the parallel here to Jacques Lacan's 1954 *Seminar 2*, where he engages cybernetics and parallels experimental epistemology, asking what it would mean for a homeostat or a circuit to have a subject.[27] He does not ask whether we are circuits (or even worse, "just circuit"), but what we would experience if we were caught in this or that particular circuit. In reading *Seminar 2* it is important to pay attention to messages as signals in McCulloch's sense, having both yes/no and true/false. The problem with signals, tragically for Lacan, is that "yes" does not always mean "true," and even not getting a signal also has meaning (like what happens after you say "I love you" to another person and he or she hesitates). Like each neuron, it is as if we are always already receiving messages and interpret both their reception and nonreception moment after moment. Lacan continues, "Suppose I send a telegram from here to Le Mans, with the request for Le Mans to send it back to Tours, from there to Sens [and back] to Paris, and so on indefinitely. What's needed is that when I reach the tail of my message, the head should not yet have arrived back. The message must have time to turn around. It turns quickly, it doesn't stop turning, it turns around in circles."[28]

Time and delays are fundamental to enacting this circuit. *Seminar 2* is full of discussions about how every hesitation can itself be a message, from experiments in haste to "The Purloined Letter" as a model of subjects moved and moving around circuits.[29] Not paying attention to time is the strangest thing today about circuits that are supposed to represent the mind or brain. The subjects are never looked for inside the circuits, which is why most cognitive neuroscientists can never seem to find them.[30] Lacan concludes one of his lectures with the following statement that I think best articulates the relational notion of being in a circuit (vs. being a circuit):

> The need for repetition . . . doesn't conform much with vital adaptation. . . . Here [considering adding machines] we rediscover what I've already pointed out to you, namely that the unconscious is the discourse of the other. This discourse of the other is not the discourse of the abstract other, of the other in dyad. . . . It is the discourse of the circuit in which I am integrated. I am one of its links. It is the discourse of my father . . . in so far as my father made mistakes which I am absolutely condemned to reproduce . . . because one can't stop the chain of discourse, and it is precisely my duty to transmit it

in its absolute form to someone else. I have to put to someone else the problem of the situation of life or death in which the chances are that it is just as likely that he will falter, in such a way that this discourse produces a small circuit in which an entire family, an entire coterie, an entire camp, an entire nation or half the world will be caught. The circular form of speech which is just at the limit between sense and non-sense, which is problematic.[31]

## Lively Flow Diagrams

*The analysis of the operation of a machine using two-indication elements and signals can be conveniently expressed in terms of a diagrammatic notation introduced, in this context, by von Neumann and extended by Turing. This was adapted from a notation used by Pitts and McCulloch.*
—Douglas R. Hartree, *Calculating Instruments and Machines*
(Urbana: University of Illinois Press, 1949)

ACCORDING TO THE few histories of programming that attend to flow charts, Herman Heinc Goldstine and John von Neumann developed the concept ex nihilo for their 1947 *Planning and Coding of Problems for an Electronic Computing Instrument*, with the help of Adele Goldstine.[32] They drew on McCulloch and Pitts's neurons as well as conventions from engineering and brain and factory diagrams, and fed them back through common engineering and chemical engineering schematics and flows (also called flow diagrams). But as S. J. Morris and O. C. Gotel observe, "the new flow diagrams of Goldstine and von Neumann altered the basic definition in two fundamental ways":[33] the flow diagram's object was abstract rather than material, and it assumed a staged process that attends to the logic of iterative processes.

Goldstine and von Neumann were in the middle of inventing the modern computer; they were trying to imagine "memory" during iterative routines and were having a tough time of it. Memory—as in RAM (random access memory), as in stored programs—was named after human memory, but this analogy didn't really help explain how it worked. Memories are not straightforward in humans, nor are they in machines.

THE PROBLEM WAS that the stored program could overwrite itself, and then, when it came back to the same place, the program would be different. So what was the "it" that came back? They named "it" C, for control. C was the control organ that executed and was executed by the memory organ. They used the example of a Turing machine: C is the "head," but one subject to and subjecting the programs to change. C was the time-place attention of the computer; it subjected the computer to its time of succession and was itself timed by the internal clock. Control was the lively subject of the computer:

> To sum up: C will, in general, not scan the coded sequence of instructions linearly. It may jump occasionally forward or backward, omitting (for the time being, but probably not permanently) some parts of the sequence, and going repeatedly through others. It may modify some parts of the sequence while obeying the instructions in another part of the sequence. Thus when it scans a part of the sequence several times, it may actually find a different set of instructions there at each passage. All these displacements and modifications may be conditional upon the nature of intermediate results obtained by the machine itself in the course of this procedure. *Hence, it will not be possible in general to foresee in advance and completely the actual course of C,* its character and the sequence of its omissions on one hand and of its multiple passages over the same place on the other, *as well as the actual instructions it finds along this course,* and their changes through various successive occasions at the same place, if that place is multiply traversed by the course of C. *These circumstances develop in their actually assumed forms only during the process (the calculation) itself, i.e. while C actually runs through its gradually unfolding course.*[34]

Time *in* the machine algorithm needed to be engaged with. The key in dealing with it was understanding the necessity of discrete time within diagrams, such that the diagrams themselves did not so much *represent* the code as immanently *enact* a process that could leave a trace so it could be reenacted in the software. A process philosophy of immanence and duration pervades their descriptions as they come to realize that the diagrams themselves represent new forms of life they have to relate to:

> Thus *the relation of the coded instruction sequence to the mathematically conceived procedure of (numerical) solution is not a statical one, that of a translation, but highly dynamical: A coded order stands not simply for its present contents at its present location, but more fully for any succession of passages of C through it*, in connection with any succession of modified contents to be found by C there, all of this being determined by all other orders of the sequence (in conjunction with the one now under consideration). This entire, potentially very involved, interplay of interactions evolves successively while C runs through the operations controlled and directed by these continuously changing instructions.[35]

In the flow diagrams of Goldstine and von Neumann however, computers were *not* theorized as models of human brains, nor human brains as models of computers; rather brain and computer and diagrams were all different species of *thought*, in this case inductive. These flow diagrams were lively, in a literal manner; they could grow and learn. Control set them apart from all previous diagrammatic types (fig. 8.3).

According to Goldstine, "In the spring of that year von Neumann and I evolved an exceedingly crude sort of geometrical drawing to indicate in rough fashion the iterative nature of an induction. At first this was intended as a sort of tentative aid to us in programming. Then that summer I became convinced that this type of flow *diagram*, as we named it, could be used as a logically complete and precise notation for expressing a mathematical problem and that indeed this was essential to the task of pro-

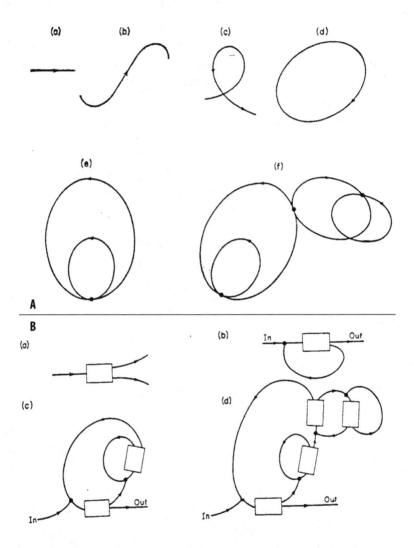

**Figures 8.3a and 8.3b**   Goldstine and von Neumann's flow diagrams. These lines with arrows in them, and then boxes, represent the dynamic, iterative processing of code inside a computer.

Source: Goldstine and von Neumann, *Planning and Coding*, figures 7.1 and 7.2, pp. 6–7.

gramming."[36] The method turned out to be a way of planning a schematic of the course of C through the code. Thus, faced with a coding problem, the procedure they suggest for the programmer is to evolve a diagram. Like the mystic writing pad that is never the same but appears new, flow diagrams and other paper tools were created to solve the impossibility of representing a process that changed itself. The diagrams showed the various states of C, its growth and its trails, and its traps (fig. 8.4). In a manner similar to the concept of acceleration, the flow diagrams captured the form of the changing memory, thus relating the beginning to the desired end of the program: "Since coding is not a static process of translation, but rather the technique of providing a dynamic background to control the automatic evolution of a meaning, it has to be viewed as a logical problem and one that represents a new branch of formal logics."[37]

The diagrams were invented to express, in mechanical form, the problem of reflexive algorithms, forms that change themselves, forms that could learn. Goldstine, von Neumann, Adele Goldstine, and Arthur W. Burkes worked it out iteratively in large measure by coding hundreds of problems included in the book.[38] Distributed in two parts, this followed up *Preliminary Discussion of the Logical Design of an Electronic Computing Instrument* by Burks, Goldstine, and von Neumann.[39] This second part was "intended to give an account of our methods of coding and of the philosophy which governs it"; they wanted to discuss the "nature of coding per se and in doing this lay down a modus operandi for handling specific problems."[40] The book then provides a series of diagrams with extensive textual description that justifies each step.

As they make clear in their text many times, the advantages of this method of diagramming are many. Primarily the diagrams provide an "economy of analysis." Errors can be traced, interventions and corrections can be proposed, and their results and consequences can be read off immediately.[41] This is effected by the fact that the diagram operates on a different level than the actual

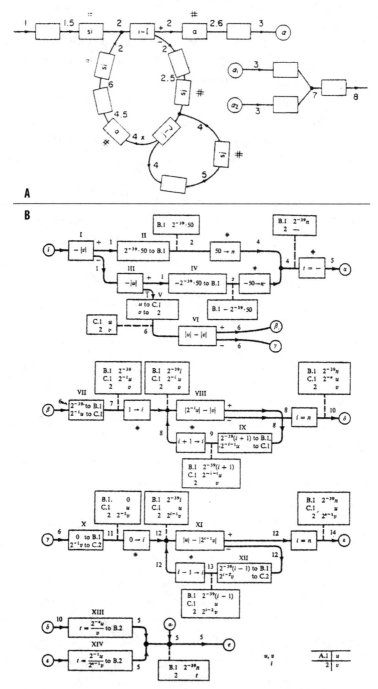

**Figures 8.4a and 8.4b**   Goldstine and von Neumann's coding problems. These diagrams demonstrate the simple and then more complicated techniques of bookkeeping used to keep track of the iterations of code as the computer was running.

Source: Goldstine and von Neumann, *Planning and Coding*, figures 7.7 and 8.2, pp. 12, 38.

code: iterations and loops do not usually affect the diagram, and hence the dynamism and evolving meaning are more or less contained. They note that the flow diagrams are designed to remain unchanged, but even that is not unqualifiedly true as there are places where one has to erase and redraw lines to keep track of C. Here the diagram is plastic in relation to the course of time (figs. 8.5a, 8.5b). They explicitly note that the flexibility of their system is that these corrections and modifications "can almost always be applied at any stage of the process without throwing out of gear the procedure of drawing the diagram, and in particular without creating a necessity of 'starting all over again.'"[42] These factors are crucial in understanding the technique's power as it migrates to other fields.[43]

This book was widely distributed and placed in the public domain from the moment of its printing. It was widely read and iterated. Goldstine and von Neumann's specific manner of flow diagramming was quite involved and often counterintuitive, despite being logically rigorous. They were never used again. Nonetheless the term *flow diagram* and the technique of representing the flow of control as a network of instructional boxes connected by arrows flourished and quickly became ubiquitous in computer programming. Just two years after its publication, a conference at Cambridge University Mathematical Laboratory called "High-Speed Automatic Calculating Machines" on June 22–25, 1949, included many kinds of flow diagrams and featured Turing apologizing that no one knew exactly how to read or write them; each person in fact developed a different mode of diagramming.[44] Here we can note two more things about flow diagrams: in addition to being extremely flexible in operation, they are adaptable in convention and style, and they are immediately comprehensible.

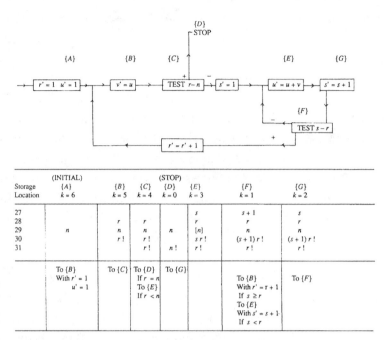

Figure 8.5a  Turing's flow diagram draws upon Goldstine and von Neumann's diagrammatic conventions but greatly simplifies the bookkeeping, allowing the diagram to now remain static with respect to time.
Source: Alan Turing, "Checking a Large Routine," in *The Early British Computer Conferences*, ed. Michael R. Williams and Martin Campbell-Kelly (1949; Cambridge, MA: MIT Press, Tomash, 1989) , 70–72.

## Plans and the Structure of Behavior

SOME OF THE first digital flow charts taken up by psychology happened via the book *Plans and the Structure of Behavior*. In 1958 two psychologists, George Miller and Eugene Galanter, and neurophysiologist Karl Pribram at the Center for Advanced Studies in Behavioral Science at Stanford attempted to "discover whether cybernetic ideas [had] any relevance for psychology."[45] They also had a pile of materials that Miller had just obtained from one of the first computer programming summer schools, run by Allan Newell, J. C. Shaw, and Herbert Simon at a Research Training Institute on the Simulation of Cognitive Processes, at RAND Corporation in July 1958.

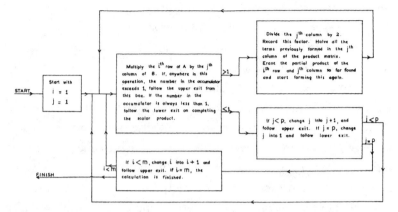

**Figure 8.5b**   Noble's diagram is remarkable for how redundant it appears, with lines having multiple arrows and boxes explaining exactly what must be done, including how to exit, and with the exits also marked. This excess of instruction indicates how new this technique of digital stepping through diagrams was at the time.
Source: B. Noble "The Control of Magnitudes of Numbers in Digital Computing Machines with a Fixed Binary Point," in *The Early British Computer Conferences*, 54–59.

THE AIM OF the book is to get past behaviorism's bias against cognitivism, the idea that an organism has ideas, something "in its head," as it were, a picture of itself and its universe. At the same time, the authors think that even if we admit to some "ghostly inner somethings," as they call them, cognitive insights are still missing a crucial mediator: motivation or will. They cite a critic (Guthrie) arguing that "[cognitivists] are so concerned with what goes on in the rat's mind, that they neglect to predict what it will do." And they continue, "It is so transparently clear to them that if a hungry rat knows where to find food—if he has a cognitive map with the food box located on it—he will go there and eat. What more is there to explain? The answer, of course, is that a great deal is left to be explained. The gap from knowledge to action looks smaller than the gap from stimulus to action—yet the gap is there, indefinitely large." Note the striking persistence here: at the core of what will become cognitive psychology is posed an inverted problem—not *What is cognition?* or *How might it go wrong?* but *How does cognition do anything? How does the brain act, even given a rationale?* This is a tragic problem for them, partly because it requires returning to the concept of the will, going back to William James, who at least thematized it as "ideo-motor action." But even that "helps us not in the least. The bridge James gives us between the ideo and the motor is nothing but a hyphen. There seems to be no alternative but to strike out into the vacuum on our own." Temptingly they make clear that the answer will not be simple: "The present book is largely the record of prolonged—and frequently violent—conversation about how that vacuum might be filled."[46]

This vacuum, into which motivation and will must be accounted for, is fascinating and violent precisely because it concerns the rational actor as problem and not solution. How does reason come to act? The paradox that the authors will contemplate at the heart of calculators even as complex as computers is that programs run; they do not reflect. The calculative program is not necessarily rational at all, and therefore there is a deep problem in the notion of "instrumental action."

The solution, as the book's title indicates, is grounded on the notion of *plans*, which are simultaneously complete descriptions of behavior and the way organisms such as humans actually make their way in the world: "A Plan is, for an organism, essentially the same as a program for a computer."[47] But, as in Goldstine and von Neumann's discovery of flow diagramming, the actual following of the plan or program involves a complex evolution of meaning. In the scheme of Miller et al., each organism has a set of plans and an Image, being "all the accumulated, organized knowledge that the organism has about itself and its world. The Image consists of a great deal more than imagery, of course."[48] Plans are developed in order to make one's way through the world as seen in the Image. These plans in turn modify the image, which in turn generates new plans.

THE AUTHORS BEGIN their analysis of plans by considering the concept of the reflex arc, stimulus response, and the cybernetic concepts of feedback and homeostasis. None of these presents the complexity that they feel is needed to explain real reflexes or human behavior. But they do not actually make an argument. Instead they draw a diagram (figs. 8.6a, 8.6b).[49]

There are only five diagrams in the entire book, and they appear within ten pages of each other at the beginning. They begin with the Test-Operate-Test-Exit unit, or TOTE. Although they initially present the diagram to illustrate the cybernetic feedback device, or servomechanism, they then take the diagram as central: "If we think of the Test-Operate-Test-Exit unit—for convenience, we shall call it a TOTE unit—as we do of the reflex arc, in purely anatomical terms, it may describe reflexes, but little else. That is to say, the reflex should be recognized as only one of many possible actualizations of a TOTE pattern. The next task is to generalize the TOTE unit so that it will be useful in a majority—hopefully, in all—of the behavioral descriptions we shall need to make." Thus they explicitly want the diagram to represent, potentially, all behavioral and cognitive descriptions. They accomplish this with an ingenious bait-and-switch rhetoric, asking, "Consider what the arrows in Figure 1 might represent. What could flow along them from one box to another? We shall discuss three alternatives: energy, information, and control." I call this a bait-and-switch because they ask a question I have not seen repeated in scholarly or scientific texts. They don't ask *what* do the arrows mean? Or how should one interpret them? But what *could* the arrows represent? They then act as if there is a choice to be made among three alternatives. And they proceed to analyze the same diagram from each perspective. In the history of flowcharts and flow diagrams, these three considerations are worth quoting at length:

> If we think of energy—neural impulses, for example—flowing from one place to another over the arrows, then the arrows

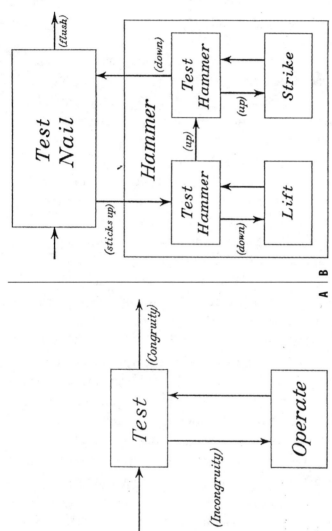

**Figures 8.6a and 8.6b**    Miller, Galanter, and Pribram's Test-Operate-Test-Exit Unit (TOTE) and Hierarchical Plan for Hammering Nails. This simple TOTE diagram served as the basis for imaging how the brain as computer program might build up simple scripts into more complicated ones. The authors were intrigued with the power of this diagram as an explanation of human behavior and also the challenges it presented if one were in the diagram following along the arrows.

Source: Miller et al., *Plans and the Structure of Behavior*, figures 1 and 5, pp. 26, 34.

must correspond to recognizable physical structures—neurons, in the example chosen. As a diagram of energy flow over discrete pathways, therefore, the TOTE unit . . . might represent a simple reflex. Or . . . a servomechanism.

There is, however, a second level of abstraction that psychologists usually prefer. We can think of *information* as flowing from one place to another over the arrows . . . as the transmission of correlation over the arrows. . . . In that case, we are concerned not with the particular structures or kinds of energy that are involved in producing the correlation but only with the fact that *events at the two ends of the arrow are correlated* . . . exactly what [psychologists] mean when they draw an arrow leading from Stimulus to Response in their S-R diagrams or when they define a reflex as a correlation between S and R but refuse to talk about the neurological basis for that correlation.

A third level of abstraction . . . is the notion that what flows over the arrows . . . is an intangible something called *control*. Or perhaps we should say that the arrow indicates only succession. This concept appears most frequently in the discussion of computing machines, where the control of the machine's operations passes from one instruction to another, successively, as the machine proceeds to execute the list of instructions that comprise the program it has been given. But the idea is certainly not limited to computers.[50]

They provide the example of looking up authors in an index.

Each of these three alternatives is semiotically rich in descriptive value. Each is a distinctly different way of interpreting the diagram, and each creates a different set of entailments. But now they present their revolution: "When [the TOTE diagram] is used in the discussion of a simple reflex it represents all three levels of description simultaneously. When it is used to describe more complex activities, however, we may want to consider only the transfer of information and control or in many instances only the transfer of control. In all cases, however, the existence of a

TOTE should indicate that an organizing, coordinating unit has been established, that a Plan is available."[51] What fascinates me most about this analysis is that it takes the ambiguity for granted as a necessary entailment of the flow diagram. Without further instructions on how *not* to interpret such a diagram, it will imply all three kinds of flow, and, they indicate, in the case of a simple reflex, it does.

Here, in a nutshell, is a key to understanding the power and persuasiveness of the flow diagrams sketched by the researchers in my opening story. The same diagram, taken to represent a reading disorder, can be read by the neuroanatomist in terms of anatomy, by the psychiatrist in terms of information and neurotransmitters, and by the literacy therapist in terms of control. Here also we can see a particular kind of diagrammatic equivalence, simultaneously describing the brain, the mind, and the computer.

But the authors of this diagram don't yet know how to identify with this image. The space of existential conflict in *Plans* is severe. Control, or C in Goldstine and von Neumann, is what hangs them up: the subject *in* the circuit versus the subject *of* them. If we are such circuits, Miller et al. are forced to wonder, how do we account for motivation? The compulsion to follow the arrows, to repeat indefinitely, seems insane. They know that their system seems too crazy, psychotic even, yet they are also completely seduced by it. They spend chapter after chapter trying to follow the arrows, but without settling the question of where the subject lives they go repeatedly out of their minds.

They say that what we mean when we talk about *intentions* are "the uncompleted parts of a Plan whose execution has begun." No problem—but we seem to be in the middle of things already. So how does Planning begin, since it can't be with another, meta-Plan? "Why should any Plan ever be executed[?] . . . Plans are executed because people are alive. This is not a facetious statement, for so long as people are behaving, some Plan or other must be executed. The question thus moves from why Plans are executed to a concern for which Plans are executed." They continue, noting that the "obvious is sometimes hard to see." "The fundamental, underlying banality, of course, is the fact that once a biological machine starts to run, it keeps running twenty-four hours a day until it dies. The dynamic 'motor' that pushes our behavior along its planned grooves . . . is located in the nature of life itself. As William James says so clearly, the stream of thought can never stop flowing."[52] We are always already inside of a Plan. Consciousness is already ongoing.

I'm dwelling on this setup because it points out a number of key properties regarding the idea of programs in the brain: (1) that having them doesn't account for either running them or choosing which one to run, and (2) that there is a funny obsession with "getting things done" with Plans. They often give office examples: filing, going to the dentist, and so on.

This leads directly to the fact that humans so often screw up and therefore a chapter on remembering: "How could anyone forget what he intends to do?" The answer is that the Plan doesn't have a Plan, and the brain doesn't have a brain. Interrupts happen. To which they consider Freud's repression and "the man whose appointment book is destroyed through no fault of his own," calling for "more research on the way people use external aids as memory devices." They further comment, "Remembering the Plan is most difficult when we try to do it without external crutches."[53] In other words, they are attentive to the lived temporality of memory in their lives.

As the book progresses, the authors become even more aware of the strangeness of placing Plans at the core of their notion of human nature. They defend it, however, as an "indispensable aspect of the human mind," but they note, "Nonetheless, a cautious reader should not overlook the American origins of this book on the psychology of Plans."[54] This is to say that they are taking their own notions very seriously. The fact that flow diagrams in computers have their origins in war and operations research, whose origins are in mapping factory floors, flows, and processes should not be taken lightly when imagining that somehow our brains are actually organized like factories or computers.

Much of the book is an inventory of key aspects of the diagram that seem to run into trouble when we think of actual social humans. How to account for language, for instance, or "Plans for speaking," as they call it? In their fight with behaviorists, who reduce language to speedy-efficient behavior coordination, they feel compelled to come up with some "new and important psychological process that is introduced by language that cannot be thought of as a technique to increase the efficiency of inarticulate processes."[55]

Why not death? "Man is the only animal that knows he is going to die. . . . A man can make death a part of his Plans, as when he buys life insurance policies or draws a will, or he can even devise and execute a Plan that intends his own death. Indeed, we occasionally hear it said that a truly rational man would commit suicide immediately. But whatever one's evaluation of life . . . "[56]

Plans now have two striking features: On one side they need the joy of life to set them in motion, and on the other the mere fact of planning one's life as such means planning for death. There is a chapter, for instance, on "relinquishing the Plan" where they meditate on the fact that if the goal of a Plan is to complete it, then what? Can plans as a whole stop? And if stopped, can they start again? How does one have a metaplan for when one is "Planless"? "To find oneself without any Plan at all is a serious matter. Interpreted literally, it is impossible, for complete planless-ness must be equivalent to death."[57]

The authors are in a quandary. They realize that if one is in a Plan, there is no way to know whether the Plan one is running is one's own or another's. Is fanaticism a Plan that can't be checked? And they speculate that psychosis might be following an old Plan that no longer applies (and not knowing it): "One possible reaction is to reinstate the old Plan in spite of the fact that it is no longer relevant or feasible, to continue to develop it, transform it, and execute it despite its inadequacy. In its most extreme form, this is the paranoid reaction." The chapter on relinquishing Plans is obsessed with hypnosis and its implications for our plans, on knowing whether we are living our own plans or someone else's, such as our parents': "One of the seven wonders of psychology is that so striking a phenomenon as hypnosis has been neglected. . . . The hypnotized person is not really doing anything different, with this exception: the voice he listens to for his Plan is not his own, but the hypnotist's. The subject gives up his inner speech to the hypnotist."[58]

The weird wonderfulness of this book is that by placing themselves inside of circuits, Miller et al. ponder a life lived *within* logical circuits. Their careful following along the paths of plans leads them to one existential crisis and insight after another: "The person who makes his life's Plans in terms of concrete and specific objectives, in terms of 'goals,' invites the disaster of planlessness. . . . The problem is to sustain life, to formulate enduring Plans, not to terminate living and planning as if they were tasks that had to be finished. This simple point has been overlooked by many psychologists who seem to take it for granted that all behavior must be oriented toward explicit goals."[59]

### Experimental Epistemology or Artificial Neuroses

THIS PRACTICE OF experimental epistemology continued throughout the latter half of the 1950s and through the 1960s, during which various projects that were then called artificial intelligence were developed but focused instead on computer simulation or models of personality.[60] Rather than intelligence, however, these projects attempted to simulate irrationality, frustration, stupidity, neuroses, and paranoia. Imagine ELIZA and you have a sense of the complexity of these programs, which were often less than one hundred lines long. Each of the programs functioned like an existential device.

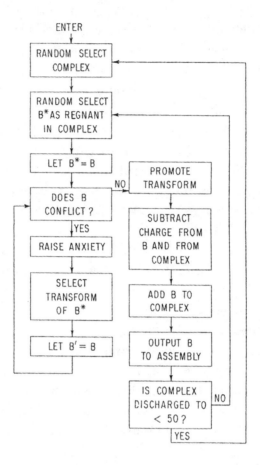

**Figure 8.7** Kenneth Mark Colby's simulation of a neurotic process. This diagram served as a demonstration of the logical nature of a neurosis. By following from box to box, a subject could turn "I hate my father" into "I hate my brother."
Source: Kenneth Mark Colby, "Computer Simulation of a Neurotic Process," in *Computer Simulation of Personality: Frontier of Psychological Research*, ed. Silvan Samuel Tomkins and S. Messick (New York: Wiley, 1963), 171.

Kenneth Mark Colby wrote a short program that would respond to demands neurotically. Given two conflicting demands, such as "I hate my father" and "I love my father," the program would return a psycholinguistic variant: "I hate my brother." The point of the exercise was to demonstrate that neuroses like this, to the extent that they could be reliably modeled, quite possibly shared this algorithmic structure, and therapy could make use of this (fig. 8.7).

Colby's second program is more famous because it is to date the only program that rigorously passed the Turing test. Called PARRY, the program was designed to simulate a paranoid individual. If you typed "What does your father do?," it would reply, "Why do you want to know about my father?" Sounds ludicrous? Like a trick? It is, and yet that is the point. When a transcript of a psychiatrist interviewing PARRY was mixed up with transcripts of interviews with real paranoid schizophrenics, a panel of psychiatrists was unable to detect the difference. Apparently the simple, mechanical tricks of the program were parallel to the mechanisms by which many paranoids constructed their answers (fig. 8.8). We see a trick; they live inside its temporality. That is how the diagram and the program function like existential devices.

A final example: Robert Abelson wrote a program whose goal was to take a fact and offer a right-wing interpretation of it; for example, if you told the program there were movements in North Korea, it would reply that the communists were behind it (fig. 8.9). Abelson's point was very explicit: "to simulate responses to foreign policy questions by a right-wing ideologue, such as Barry Goldwater."[61] He called it the "Ideology Machine." To the extent that the simple program could mimic Goldwater's responses, right-wing ideology could be shown to be mechanical.[62]

For Abelson, Colby, and others, the simulations were both simple and powerful. In many contemporary histories of artificial intelligence, these receive little if any attention.[63] To the extent they do (ELIZA and to a lesser extent PARRY), they are treated as cute "frauds" or tricks, ultimately failing to "adequately" account for human intelligence.[64] I think that this definition misses their uses, which were much closer to those of McCulloch's existential devices, in which one learns something about paranoia in tandem with being able to simulate a part of

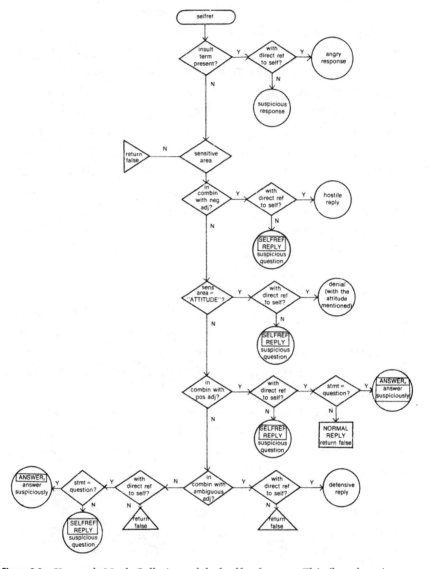

**Figure 8.8** Kenneth Mark Colby's model of self-reference. This flowchart is part of a larger program for simulating a paranoid subject. As with neuroses, Colby was interested in the logic of irrational psychic structures, patho-logics.
Source: Kenneth Mark Colby, *Artificial Paranoia: A Computer Simulation of Paranoid Processes* (New York: Pergamon Press, 1975), 61.

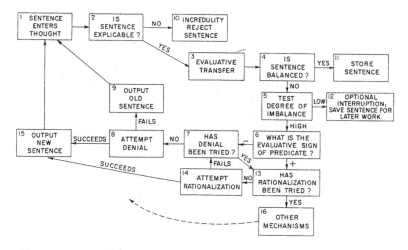

**Figure 8.9** Robert Abelson's "Flow Chart for Cognitive Balancing." Abelson designed a simulation of "hot" cognition, of the logically emotional interpretive strategies of a right-wing ideologue, such as Barry Goldwater. Source: Abelson, "Computer Simulation of 'Hot' Cognition," 289.

its linguistic logic. And this happens interactively in getting the program right. The program for these simulators was meaningful only when run; Margaret A. Boden touches on this issue. But it is the intersubjective nature of the run that is the focus. The *program* is not the object to be compared with a paranoid person; *a run of the program* interviewed by the psychiatrist is the proper comparison. This is a process approach to crafting artificial intelligence (AI) that some contemporary research in science, technology, and society recognizes.[65]

One of my aims in tracing the lineages of these diagrams is to analyze how the same diagrams come to be used in very different ways, in one case to highlight a problem, in another to bypass that same problem. For example, in Allen Newell's combined praise and critical discussion in 1972 of cognitive psychology's use of flow diagrams and binaries, he notes that flow diagrams in cognitive psychology are advances over verbal theories by "asserting the existence of an entire set of processing stages or components and some orderings between them." But he adds, "They do not model the control structure"—meaning a programming language such as FORTRAN that constrains the kinds of operations and limitations of the machine. He then provides an example, one of the foundational cognitive psychological works on memory by Richard C. Atkinson and Richard M. Shiffrin in 1968 (fig. 8.10a):

> The model of memory is there all right, and is applied to a number of tasks with quantitative precision. However, the control structure is completely absent and is used as a deus ex machina to concoct separate models for each task. Criticism is not directed at that justly influential piece of work. But it does illustrate well the current state of the theoretical art. As long as the control structure—the glue—is missing, so long it will be possible to suggest in indefinite sequence of alternative possibilities for how a given task was performed, hence to keep theoretical issues from becoming settled.[66]

Subsequent versions of the same diagram show a successive repression of even this ghostly control structure, first to dotted lines, then erased altogether (fig. 8.10b). The result is a static map of modules rather than a dynamic model or existential device. Newell is essentially arguing that there is a lack of existential commitment to the running computer in time. He nonetheless also argued that the ambiguity in the diagrams was still productive, of new research papers if nothing else![67]

A similar paradox haunted the last chapter of Miller et al.'s *Plans and the Structure of Behavior*, where the authors at last take on the problem of the brain's relation to psychology, which they have explicitly avoided up to then. There was a fundamental disagreement between Miller and Galanter, who believed that cognition and the brain are computer-like, and Pribram, who believed that they are more biological.[68] They reveal in a footnote that they argued severely and without resolution regarding this question, almost to the point of not writing the book: "The arguments revolved around a three-way analogy: The relation of a Plan to

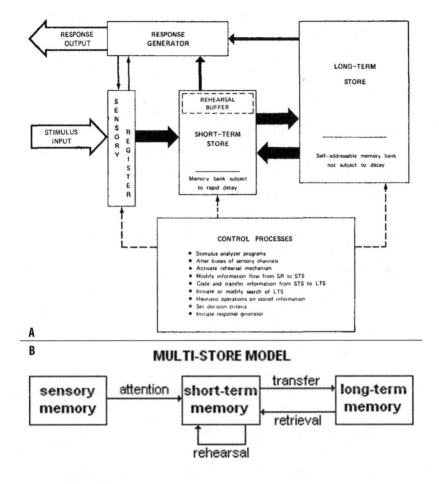

**Figures 8.10a and 8.10b** Two iterations of the Atkinson-Shiffrin memory diagram, in neither of which is any notion of timing present. In the first, control processes are represented by a "Control Processes" box connecting with dotted lines to the other boxes. In the second, used in Wikipedia to represent the same diagram, the control processes are removed entirely. Sources: Richard M. Shiffrin and Richard C. Atkinson, "Storage and Retrieval Processes in Long-term Memory," *Psychological Review* 76, no. 2 (1969): 179; "Atkinson-Shiffrin Memory Model," Wikipedia, https://en.wikipedia.org/wiki/Atkinson-Shiffrin_memory_model. The original paper does not have this diagram, but it is nonetheless attributed to Richard C. Atkinson and Richard M. Shiffrin, "Human Memory: A Proposed System and Its Control Processes," in *The Psychology of Learning and Motivation*, ed. K. W. Spence and J. T. Spence (New York: Academic Press, 1968), 2: 89–195.

the mind is analogous to the relation of a program to a computer, and both are analogous to the relation of X to the brain. Question: What is X? Is it possible to locate parts of the brain that correspond, however crudely, to these parts of a computer?" The problem with this analogy, they reveal, is that it is exceedingly open-ended. Each solution that fits the available data is not only promising and suggestive of new research; it is also ad hoc: "After several months of discussion, the present authors were almost (but not quite) convinced that you could put the names of parts of the brain on slips of paper, scramble them up, draw two at random, assign them in either order to serve either as the memory or as the processing unit, and you would be able to interpret *some* evidence *somehow* as proof that you were right. . . . It is wonderful to see how these analogies can blossom when they are given a little affection."[69]

In sum, the function of the careful explication of the triple referent of the flow diagram was not to encourage better diagrams or more analytic readings but in fact to enable and authorize productively polyvalent diagrams. The triple reading thus allows the diagrammer and the reader to *act as if* brains, behavior, and cognition were functionally organized like diagrams and like computers. Although the diagrams impose some constraints by means of boxes and the direction of the arrows between them, they primarily function as analogies and models that make the kind of conscious behaviorism proposed by them seem understandable.[70] They provide a space for cooperation without agreement, in James R. Griesemer and William C. Wimsatt's terms, "conceptual maps . . . which fix conceptual environments for a variety of problems."[71]

### Summary: Plastic Diagrams x 2

IN THIS PAPER I have all too rapidly explored the variety of meanings that flowcharts have had since their digital inception. For

Goldstine and von Neumann they were flexible, erasable paper tools to keep track of time in running processes and to thereby notice the incredible liveliness of mechanisms. For McCulloch these existential devices demonstrated (among other things) that if we are mechanisms like these, we will be neurotic, psychotic, dogmatic, and so on. We will be neurotic in our parts, logically, so to speak, while in other parts logically paranoid, and in still others obsessive, attentive, abstracting, willful, and so on.

While flowcharts started out this way for Miller, Galanter, and Pribram, as plans that logically ran us (rather than our running them), the yearlong conversation they had transformed the diagrams into plastic spaces of productive disagreement and ambiguity over the correct analogies. Newell's 1972 critique of the state of the field acknowledged the transition, so to speak, from temporal to spatial model, from running computers to black boxes (with more boxes inside). The arrows that mattered so much for Miller, Galanter, and Pribram vanished altogether into simple lines.

Telling a story of flowcharts this way makes them akin to the story of Feynman's diagrams as narrated so beautifully by David Kaiser, where the specificity of their initial use mutates into an altogether different practice while maintaining visual similarity.[72] The diagrammatic conventions are therefore historically plastic, while at each moment the diagrams were being used as plastic "conceptual scaffolds" or "paper tools" in ways that are both generative and limiting.[73] Intriguingly, contemporary computer science, AI research, cognitive psychology, and neuroscience have a lively critique of flowcharting that they each call "boxology," sometimes defined as "the ostentatious display of flow-diagrams as a substitute for thought."[74]

The hypothesis that the research in this chapter generates is that diagrams do more than exercise powerful constraints on how theories develop and how experiments are imagined.[75] The evolution of diagrams is an extremely useful indicator of conceptual change, of agreement on the framing of problems (though not on solutions), and of units of thought, theorizing, and experimentation. By following their theoretical and practical entailments within and across disciplines, we can better understand the importance of both visual and metaphorical studies in the history of science and technology. By concentrating on the actual changes in diagrammatic practices of flowcharts and their semiotic use in publications, we can track how they often cement and reify the mutual implication of computers and brains.

As we turn to the plastic brain to escape the apparent limitations of the nonplastic too-computer-like brain, we may be misplacing our fears. My critical suspicion is that in the turn away from the analogy of *brain as a running computer* and the turn toward the notion of *brain as computer*, contemporary cognitive psychology and neuroscience have lost more than time. It is not so much a poor view of the brain as an impoverished view of computers. The idea that computers are not mysterious because they are logical mechanisms misses the core point of the early diagrammers: there are many kinds of running computers, many kinds of logical mechanisms in time, just as there are many kinds of running brains, and parts of brains, and parts of minds. Both have a plasticity with relation to time. The former are simple enough to build and spend time with and can potentially teach us about the different ways of being in and with the latter. This, in rapid form, is one genealogy of our sense of mechanism, brain, computer, and self. Perhaps resurrecting some of these existential devices and experimenting with epistemologies again will offer new avenues for lively thought and discussion.

# Endnotes

1 This paper was supported by an NSF Scholar's Award, "How Flowcharts Got into the Brain: Diagramming Brains, Minds and Computers Together," NSF# 0924988. Versions of it have been presented at ten venues since 1998, and my many colleagues will have to wait for the book to be properly thanked. It has also benefited greatly from brainstorming and research assistance by Melissa Salm, Tory Webster, Evan Buswell, Xan Chacko, and Carlos Andrés Barragán.

2 See Joseph Dumit, *Picturing Personhood: Brain Scans and Biomedical Identity* (Princeton, NJ: Princeton University Press, 2004); Joseph Dumit, "How (Not) to Do Things with Brain Images," in *Representation in Scientific Practice Revisited*, ed. Catelijne Coopmans, Jantet Vertesi, Michael Lynch, and Steve Woolgar (Cambridge, MA: MIT Press, 2014), 291–313.

3 The reading I'm tracing is coeval with but separate from Andrew Pickering's study of cybernetics; see Andrew Pickering, *The Cybernetic Brain: Sketches of Another Future* (Chicago: University of Chicago Press, 2010). While he attends to the notions of emergence, performance, and becoming in the lively machines that Ross Ashby, Grey Walter, and Stafford Beer built or tried to build, the diagrams that I'm following are not built but traced. Life, if I can temporarily use that word, happens in them, not as a result of them.

4 "Because each [author] showed there could be a machine that turned the trick, it were best to see them—at least from the logical point of view—as existential devices." Warren S. McCulloch, "Mysterium Iniquitatis of Sinful Man Aspiring into the Place of God," *Scientific Monthly* 80, no. 1 (1955): 35–39. While McCulloch has been studied in the history and philosophy of science, there has been less work on his diagrams. See Steve Joshua Heims, *The Cybernetics Group* (Cambridge, MA: MIT Press, 1991); N. Katherine Hayles, *How We Became Posthuman: Virtual Bodies in Cybernetics, Literature, and Informatics* (Chicago: University of Chicago Press, 1999); Lily E. Kay, "From Logical Neurons to Poetic Embodiments of Mind: Warren S. McCulloch's Project in Neuroscience," *Science in Context* 14, no. 4 (2001): 591–614; Tara H. Abraham, "Integrating Mind and Brain: Warren S. McCulloch, Cerebral Localization, and Experimental Epistemology," *Endeavour* 27 (2003): 32–36; Tara H. Abraham, "From Theory to Data: Representing Neurons in the 1940s," *Biology and Philosophy* 18, no. 3 (2003): 415–26; Tara H. Abraham, "Transcending Disciplines: Scientific Styles in Studies of the Brain in Mid-Twentieth Century America," *Studies in History and Philosophy of Science Part C: Studies in History and Philosophy of Biological and Biomedical Sciences* 43, no. 2 (2012): 552–68; Jean-Pierre Dupuy, *The Mechanization of the Mind: On the Origins of Cognitive Science* (Princeton, NJ: Princeton University Press, 2000). My work here builds on Abraham's excellent research into the origins and evolution of their diagrams (see Abraham, "From Theory to Data"). While she focuses on

their relation to anatomical neurons, I am interested in the strong existential reflections McCulloch makes throughout all of his papers and how they are deeply tied up with the way he uses and interacts with diagrams. In spirit, I'm closer to Kay's attention to his poetic embodiment of mind (Kay, "From Logical Neurons to Poetic Embodiments of Mind"). Kay and I worked together on his diagrams and sparred over these embodiments, and this paper pushes those discussions in the direction of embodiments of time.

5  Warren S. McCulloch and W. Pitts, "A Logical Calculus of Ideas Immanent in Nervous Activity," *Bulletin of Mathematical Biophysics* 5 (1943): 115–33.

6  See Rudolf Carnap, *The Logical Syntax of Language* (New York: Harcourt, Brace, 1938). See also Kay, "From Logical Neurons to Poetic Embodiments of Mind" Michael A. Arbib, "Warren McCulloch's Search for the Logic of the Nervous System," *Perspectives in Biology and Medicine* 43, no. 2 (2000): 193–216.

7  McCulloch and Pitts, "A Logical Calculus of Ideas Immanent in Nervous Activity." See Tara H. Abraham, "(Physio)logical Circuits: The Intellectual Origins of the McCulloch-Pitts Neural Networks," *Journal of the History of Behavioral Sciences* 38, no. 1 (2002): 3–25, for a detailed account of the history of the ideas and collaborations leading up to this paper. I am emphasizing a very different aspect of the paper than Abraham is concerned with: the conclusions and existential consequences of them rather than the biological precursors.

8  McCulloch and Pitts, "A Logical Calculus of Ideas Immanent in Nervous Activity," 121.

9  Arbib, "Warren McCulloch's Search for the Logic of the Nervous System," 199. On time in neurons, see also Lily E. Kay, "Cybernetics, Information, Life: The Emergence of Scriptural Representations of Heredity," *Configurations* 5, no. 1 (1997): 23–91.

10  Pitts had figured out how to take Carnap's propositional calculus and put it in time, rupturing the equivalence function and introducing temporal propositions; see Arbib, "Warren McCulloch's Search for the Logic of the Nervous System."

11  Abraham, "From Theory to Data," 425.

12  Warren S. McCulloch, Rook McCulloch, and Heinz von Foerster, *Collected Works of Warren S. McCulloch* (Salinas, CA: Intersystems, 1989), 163.

13  In an article titled "Neuroexistentialism" I follow out this implication for the sensorium: every signal is both on or off and also true or false, but truth depends on the subject for whom it would be true. Each part of a circuit is therefore in a sensory/hallucinatory relation to each other part; each part can be seen as a subject. See Joseph Dumit,

"Neuroexistentialism," in *Embodied Experience, Technology, and Contemporary Art*, ed. Caroline A. Jones (Cambridge, MA: MIT Press, 2006) , 182–89.

14  Abraham, "Integrating Mind and Brain."

15  McCulloch and Pitts, "A Logical Calculus of Ideas Immanent in Nervous Activity," 131.

16  Ibid., 132.

17  Warren S. McCulloch, "A Heterarchy of Values Determined by the Topology of Nervous Nets," *Bulletin of Mathematical Biophysics* 7, no. 2 (1945): 89–93.

18  Warren S. McCulloch, "Brain and Behavior," in *Comparative Psychology Monograph*, vol. 20, ed. Ward C. Halstead (Berkeley: University of California Press, 1950), 39–50, reprinted as "Machines That Think and Want," in Warren S. McCulloch, *Embodiments of Mind* (Cambridge, MA: MIT Press, 1965), 307–18.

19  Norbert Wiener, *Cybernetics, or, Control and Communication in the Animal and the Machine* (New York: Wiley & Sons, 1948), 143.

20  McCulloch, "Brain and Behavior," 41.

21  Warren S. McCulloch, "The Brain as a Computing Machine," *Electrical Engineering* 68, no. 6 (1949): 492–97.

22  In this I differ from both Orit Halpern, *Beautiful Data: A History of Vision and Reason since 1945* (Durham, NC: Duke University Press, 2015) and Abraham, "Integrating Mind and Brain." Ironically for most other readers, one of McCulloch's main targets here is Claude Shannon (who most think of as the communication engineer par excellence). This is because Shannon's notion of messages "need not have semantic content" and because Shannon does not make *time* a significant variable. But that is a story for a different paper. For an excellent perspective on this point, see Gualtiero Piccinini and Andrea Scarantino, "Computation vs. Information Processing: Why Their Difference Matters to Cognitive Science," *Studies in History and Philosophy of Science Part A* 41, no. 3 (2010): 237–46; Kay, "Cybernetics, Information, Life."

23  Warren S. McCulloch, "How Nervous Structures Have Ideas," *American Neurological Association* 74 (1949): 10–11.

24  Umberto Eco defined a sign as "anything that can be used to lie." Umberto Eco, *A Theory of Semiotics* (Bloomington: Indiana University Press, 1976).

25  On "center of indetermination," see Gilles Deleuze, *Cinema 2: The Time-Image*, trans. Hugh Tomlinson and Robert Galeta (London: Athlone, 1989). Deleuze is modifying Bergson's use of the phrase; see Henri Bergson, *Matter and Memory*, trans. Nancy Margaret Paul and W. Scott Palmer (London: G. Allen, 1988). The concept is discussed best in David

Norman Rodowick, *Gilles Deleuze's Time Machine* (Durham, NC: Duke University Press, 1997).

[26] McCulloch, "The Brain as a Computing Machine," 496.

[27] Joseph Dumit, "Paranoid Circuits in Lacan and Artificial Personalities," forthcoming; Jacques Lacan, *The Seminar of Jacques Lacan. Book 2: The Ego in Freud's Theory and in the Technique of Psychoanalysis 1954–1955*, trans. Sylvana Tomaselli, with Notes by John Forrester (Cambridge: Cambridge University Press, 1988). For related approaches to Lacan and cybernetics, see John Johnston, *The Allure of Machinic Life: Cybernetics, Artificial Life, and the New AI* (Cambridge, MA: MIT Press, 2008); Lydia H. Liu, *The Freudian Robot: Digital Media and the Future of the Unconscious* (Chicago: University of Chicago Press, 2010); Nicolas Langlitz, "Lacan's Practice of Variable-Length Sessions and His Theory of Temporality," PhD dissertation, Institute for the History of Medicine, Freie Universität Berlin, 2004.

[28] Lacan, *The Seminar*, 88.

[29] Ibid., 89.

[30] "It's funny, this turning back on itself. It's called feedback, and it's related to the homeostat. [But] here, it's more complicated. . . . What is a message inside a machine? Something that proceeds by opening and not opening . . . by yes or no. It's something articulated, of the same order as the fundamental oppositions of the symbolic register. . . . It is always ready to give a reply and be completed by this selfsame act of replying, that is to say by ceasing to function as an isolated and closed circuit. . . . Now this comes very close to what we can conceive of as Zwang, the compulsion to repeat" (Lacan, *The Seminar*, 88–89).

[31] Ibid., 89–90.

[32] Herman Heine Goldstine and John von Neumann, *Planning and Coding of Problems for an Electronic Computing Instrument*, 3 vols. (Princeton, NJ: Institute for Advanced Study, 1947–48); Herman Heine Goldstine, *The Computer from Pascal to von Neumann* (Princeton, NJ: Princeton University Press, 1972). See also Donald E. Knuth, "Structured Programming with Go to Statements," *Computing Surveys* 6, no. 4 (1974): 261–301; S. J. Morris and O. C. Z. Gotel, "Flow Diagrams: Rise and Fall of the First Software Engineering Notation," in *Diagrammatic Representation and Inference*, ed. Dave Barker-Plummer, Richard Cox, and Nik Swoboda (Berlin: Springer, 2006) , 130–44.

[33] Morris and Gotel, "Flow Diagrams," 136–37.

[34] Goldstine and von Neumann, *Planning and Coding of Problems for an Electronic Computing Instrument*, 1: 1–2, emphasis added.

[35] Ibid., 1: 2, emphasis added.

[36] He continues, "Accordingly, I developed a first, incomplete version and began work on the paper called *Planning and Coding*. . . . Von Neumann

and I worked on this material with valuable help from Burks and my wife. Out of this was to grow not just a geometrical notation but a carefully thought out analysis of programming as a discipline. This was done in part by thinking things through logically, but also and perhaps more importantly by coding a large number of problems. Through this procedure real difficulties emerged and helped illustrate general problems that were then solved" (Goldstine, *The Computer*, 266–77).

[37] Ibid., 269.

[38] Ibid., 269.

[39] Arthur W. Burks, Herman Heine Goldstine, and John von Neumann, *Preliminary Discussion of the Logical Design of an Electronic Computing Instrument* (Princeton, NJ: Institute for Advanced Study, 1946).

[40] Goldstine, *The Computer from Pascal to von Neumann*, 80–81.

[41] See James R., Griesemer, "Must Scientific Diagrams Be Eliminable? The Case of Path Analysis," *Biology and Philosophy* 6, no. 2 (1991): 155–80.

[42] Goldstine and von Newman, *Planning and Coding of Problems for an Electronic Computing Instrument*, 1: 20.

[43] Flow diagrams that looked similar had existed for most of the twentieth century. Brains were diagrammed based on purported neural projections, but in most cases these were treated like reflexes. There was no reprogramming or rediagramming. Factories were also diagrammed as flows, often explicitly drawing inspiration from neural diagrams, or even neural evolution.

[44] Surprisingly while von Neumann is mentioned four times in *The Early British Computer Conferences*, there is no mention of *Planning and Coding of Problems for an Electronic Computing Instrument* and its diagrams; see Michael R. Williams and Martin Campbell-Kelly, eds., *The Early British Computer Conferences* (1949; Cambridge, MA: MIT Press, Tomash, 1989). The Electronic Delay Storage Automatic Calculator (EDSAC) demonstration by W. Renwick begins, "During the first day of the Conference, a demonstration was given of the new Cambridge Calculator, the E.D.S.A.C. . . . It is hoped that the following notes, flow diagrams, and annotations will render the actual routines used in the demonstration intelligible." W. Renwick, "Demonstration of the E.D.S.A.C. (Account Prepared by B. H. Worsley)," in Williams and Campbell-Kelly, *The Early British Computer Conferences*, 21–25. Another paper, presented by Dr. H. Eggink and titled "The Programming of Supersonic Nozzle Flow," also includes a "flow diagram" (placed in quotation marks) and some instructions on how to read it, as well as a lengthy text walkthrough of the process (50–53). One of the participants at that conference was Alan Turing; he presented a paper titled "Checking a Large Routine." In it he regretted, "Unfortunately, there is no coding system sufficiently generally known to justify giving the routine for this process in full, but the flow diagram given in Fig. 1, will be sufficient for illustration" (70–72).

He then describes exactly how to read the flow diagram and how to use it to verify the running code, again without any attribution, even though this seems to be based directly on Goldstine and von Neumann's "constancy intervals"; see Goldstine and von Neumann, *Planning and Coding of Problems for an Electronic Computing Instrument*.

45  George A. Miller, Eugene Galanter, and Karl H. Pribram, *Plans and the Structure of Behavior* (New York: Holt, Rinehart & Winston, 1960), 3.

46  Ibid., 9, 11–12. See also William James, *The Principles of Psychology* (New York: H. Holt, 1890).

47  They argue, "Newell, Shaw, and Simon have explicitly and systematically used the hierarchical structure of lists in their development of 'information-processing languages' that are used to program high-speed digital computers to simulate human thought processes. Their success in this direction—which the present authors find most impressive and encouraging—argues strongly for the hypothesis that a hierarchical structure is the basic form of organization in human problem-solving. Thus, we are reasonably confident that 'program' could be substituted everywhere for 'Plan' in the following pages. However, the reduction of Plans to nothing but programs is still a scientific hypothesis and is still in need of further validation. For the present, therefore, it should be less confusing if we regard a computer program that simulates certain features of an organism's behavior as a theory about the organismic Plan that generated the behavior" (Miller et al., *Plans and the Structure of Behavior*, 16).

48  Ibid., 17.

49  Thus they develop the idea of a TOTE: "The threshold, however, is only one of many different ways that the input can be tested. Moreover, the response of the effector depends upon the outcome of the test and is most conveniently conceived as an effort to modify the outcome of the test. The action is initiated by an 'incongruity' between the state of the organism and the state that is being tested for, and the action persists until the incongruity (i.e., the proximal stimulus) is removed. The general pattern of reflex action, therefore, is to test the input energies against some criteria established in the organism, to respond if the result of the test is to show an incongruity, and to continue to respond until the incongruity vanishes, at which time the reflex is terminated. Thus, there is 'feedback' from the result of the action to the testing phase, and we are confronted by a recursive loop. The simplest kind of diagram to represent this conception of reflex action—an alternative to the classical reflex arc—would have to look something like [the TOTE diagram]" (ibid., 26–27).

50  Ibid., 27, emphasis added. They continue: "As a simple example drawn from more familiar activities, imagine that you wanted to look up a particular topic in a certain book in order to see what the author had to say about it. You would open the book to the index and find the topic. Following the entry is a string of numbers. As you look up each page

reference in turn, your behavior can be described as under the control of that list of numbers, and control is transferred from one number to the next as you proceed through the list. The transfer of control could be symbolized by drawing arrows from one page number to the next, but the arrows would have a meaning quite different from the two meanings mentioned previously. Here we are not concerned with a flow of energy or transmission of information from one page number to the next but merely with the order in which the 'instructions' are executed" (28–29).

51  Just to complete the initial description: "In the following pages we shall use the TOTE as a general description of the control processes involved; the implications it may have for functional anatomy will remain more or less dormant until Chapter 14, at which point we shall indulge in some neuropsychological speculations. Until then, however, the TOTE will serve as a description at only the third, least concrete, level. In its weakest form, the TOTE asserts simply that the operations an organism performs are constantly guided by the outcomes of various tests" (ibid., 29).

52  Ibid., 61, 62, 64.

53  Ibid., 69–70.

54  Ibid., 102.

55  Ibid., 141.

56  Ibid.

57  Ibid., 112.

58  Ibid., 115, 104–5.

59  Ibid., 114.

60  See Silvan Samuel Tomkins and S. Messick, eds., *Computer Simulation of Personality: Frontier of Psychological Theory* (New York: Wiley, 1963); John C. Loehlin, *Computer Models of Personality* (New York: Random House, 1968); Michael J. Apter, *The Computer Simulation of Behaviour* (London: Hutchinson, 1970); Roger C. Schank and Kenneth Mark Colby, *Computer Models of Thought and Language* (San Francisco: W. H. Freeman, 1973).

61  Robert P. Abelson, "Computer Simulation of 'Hot' Cognition," in Tomkins and Messick, *Computer Simulation of Personality*, 277–98. See also Roger C. Schank and Ellen J. Langer, *Beliefs, Reasoning, and Decision Making: Psychologic in Honor of Bob Abelson* (Hillsdale, NJ: L. Erlbaum, 1994).

62  See Schank and Colby, *Computer Models of Thought and Language*; Schank and Langer, *Beliefs, Reasoning, and Decision Making*.

63  See Margaret A. Boden, *Mind as Machine: A History of Cognitive Science* (New York: Clarendon Press, 2006); Daniel Crevier, *AI: The Tumultuous History of the Search for Artificial Intelligence* (New York: Basic Books, 1992).

64   See Margaret A. Boden, *Computer Models of Mind: Computational Approaches in Theoretical Psychology* (Cambridge: Cambridge University Press, 1988); Alison Adam, *Artificial Knowing: Gender and the Thinking Machine* (New York: Routledge, 1998).

65   See, for example, Lucy A. Suchman, *Plans and Situated Actions: The Problem of Human-Machine Communication* (Cambridge: Cambridge University Press, 1987); Harry M. Collins, *Artificial Experts: Social Knowledge and Intelligent Machines* (Cambridge, MA: MIT Press, 1990); Lucy A. Suchman and R. H. Trigg, "Artificial Intelligence as Craftwork," in *Understanding Practice: Perspectives on Activity and Context*, ed. S. Chaiklin and J. Lave (Cambridge: Cambridge University Press, 1993), 144–78; Joseph Dumit, "Artificial Participation: An Interview with Warren Sack," in *Zeroing In on the Year 2000: The Final Edition*, ed. George E. Marcus (Chicago: University of Chicago Press, 2000), 59–87; Diana E. Forsythe, *Studying Those Who Study Us: An Anthropologist in the World of Artificial Intelligence* (Stanford: Stanford University Press, 2001); Warren Sack, "Network Aesthetics," in *Database Aesthetics*, ed. Victoria Vesna (Minneapolis: University of Minnesota Press, 2007), 183–210.

66   Allen Newell, "You Can't Play 20 Questions with Nature and Win: Projective Comments on the Papers of This Symposium," in *Visual Information Processing: Proceedings of the Eighth Annual Carnegie Symposium on Cognition, Held at the Carnegie-Mellon University, Pittsburgh, Pennsylvania, May 19, 1972*, ed. William G. Chase (New York: Academic Press, 1973), 296–97, 298. See also Richard C. Atkinson and Richard M. Shiffrin, "Human Memory: A Proposed System and Its Control Processes," in *The Psychology of Learning and Motivation: Advances in Research and Theory*, ed. K. W. Spence and J. T. Spence (New York: Academic Press, 1968), 2: 89–195.

67   For an extended account of Newell's address, see Joseph Dumit, "Plastic Neuroscience: Studying What the Brain Cares About," *Frontiers in Human Neuroscience* 8 (2014): 1–4.

68   The question that they want to address is to what extent a psychological theory should attempt to "stand up to" neurophysiology, given that "each new finding in one field 'suggests a corresponding insight in the other.' The procedure of looking back and forth between the two fields is not only ancient and honorable—it is always fun and occasionally useful" (Miller et al., *Plans and the Structure of Behavior*, 196).

69   Ibid., 197, 199.

70   On uses of diagrams like this, see Peter. J. Taylor and Ann. S. Blum, "Ecosystems as Circuits: Diagrams and the Limits of Physical Analogies," *Biology and Philosophy* 6, no. 2 (1991): 275–94; Mary B. Hesse, *Models and Analogies in Science* (Notre Dame, IN: University of Notre Dame Press, 1966).

71   James R. Griesemer and William C. Wimsatt, "Picturing Weismannism: A Case Study of Conceptual Evolution," in *What the Philosophy of Biology Is:*

*Essays Dedicated to David Hull*, ed. M. Ruse (Dordrecht: Kluwer Academic, 1989), 78.

[72] David Kaiser, *Drawing Theories Apart: The Dispersion of Feynman Diagrams in Postwar Physics* (Chicago: University of Chicago Press, 2005).

[73] On conceptual scaffolds, see Griesemer and Wimsatt, "Picturing Weismannism"; Linda R. Caporael, James R. Grisemer, and William C. Wimsatt, eds., *Developing Scaffolds in Evolution, Culture and Cognition* (Cambridge, MA: MIT Press, 2014). On paper tools, see Ursula Klein, *Experiments, Models, Paper Tools: Cultures of Organic Chemistry in the Nineteenth Century* (Stanford: Stanford University Press, 2003).

[74] The quote is ascribed to Stuart Sutherland and refers to cognitive psychology, or more generally "black-boxology"; see V. S. Ramachandran, *The Tell-Tale Brain: A Neuroscientist's Quest for What Makes Us Human* (New York: Norton, 2012), 297. For more on this critique in neuroscience, see V. S. Ramachandran and S. Blakeslee, *Phantoms in the Brain: Probing the Mysteries of the Human Mind* (New York: Quill, 1999); Alan J. Parkin, *Explorations in Cognitive Neuropsychology* (Oxford: Blackwell, 1996). On "anti-boxology" in AI, see Phoebe Sengers, "Anti-Boxology: Agent Design in Cultural Context," PhD dissertation, Carnegie Mellon University, 1998.

[75] On the history of physics, see Kaiser, *Drawing Theories Apart*; Arthur I. Miller, *Imagery in Scientific Thought: Creating Twentieth Century Physics* (Boston: Birkhauser, 1984). On biology, see Jane Maienschein, "From Presentation to Representation in E. B. Wilson's The Cell," *Biology and Philosophy* 6, no. 2 (1991): 227–54; Griesemer and Wimsatt, "Picturing Weismannism." On chemistry, see Klein, *Experiments, Models, Paper Tools*. On mathematics, see Brian Rotman, *Mathematics as Sign: Writing, Imagining, Counting. Writing Science* (Stanford: Stanford University Press, 2000). On ecology, see Taylor and Blum, "Ecosystems as Circuits."

*Katja Guenther*

# 9 Imperfect Reflections: Norms, Pathology, and Difference in Mirror Neuron Research

IN THE EARLY 1990s the neurophysiologist Giacomo Rizzolatti and his research group in the Department of Neuroscience at the University of Parma in Italy described a group of cells in the premotor cortex of monkeys that presented an unusual response pattern.[1] The cells, located in the rostral part of the inferior premotor cortex (F5 of area 6),[2] fired not only when the monkeys performed a given action (such as grasping a raisin) but also when they observed somebody else perform that same movement. In recognition of the similarity between the observed and executed movements, in 1996 Rizzolatti christened the cells "mirror neurons."[3]

Mirror neurons have since become the subject of much debate. In 2012 they were termed the "most hyped concept in neuroscience,"[4] in part because they had been marshaled to make claims about a wide range of questions: the origins of language, empathy, and forms of pathology such as autism. Due to their centrality in these fields, some commentators in more enthusiastic moments have seen in them the key to what makes us human: the neuroscientist V. S. Ramachandran argued in a 2009 TED talk that they were the "neurons that shaped civilization."[5]

The extravagance of such claims has incited the suspicion of academics from the social sciences and humanities.[6] Ruth Leys and Allan Young have criticized the tendency of mirror neuron researchers to sideline cognition in their understanding of human mental activity.[7] Mirror neuron researchers, in this argument, rely on an assumption that so-called basic emotions do "not involve 'propositional attitudes' or beliefs about the emotional objects in our world. Rather, they are rapid, phylogenetically old, automatic responses of the organism." In mirroring other people, the argument goes, we don't think about their situation but rather place ourselves "in their shoes," *feel* what it is like to be them. In this way, Leys has argued, the elevation of mirror neurons to a sort of human quintessence has tended to prioritize excessively the emotional and the immediate.[8]

These scholars have focused their analyses on the way mirror neurons have been used to explain empathy. But while the study of empathy has certainly been the most prominent of the research directions resulting from the discovery of mirror neurons, it is only one strand of a larger paradigm. Like many other neuroscientific paradigms in the contemporary period, mirror neuron research has a complex institutional ecology. Its core is the neurophysiology lab in Parma, whose most prominent members are Rizzolatti, Vittorio Gallese, Leonardo Fogassi, and Maurizio Gentilucci. Parma is, however, but one node in a transnational network of research institutes and working groups. As the institutional and scientific prestige of mirror neuron research grew, the Parma team entered into collaboration with other researchers: on the question of language Rizzolatti worked with computer scientist Michael Arbib at the University of Southern California; for empathy the Parma team drew on the imaging expertise of scientists like Marco Iacoboni at UCLA and Bruno Wicker in Marseille; for arguments about theory of mind they collaborated with analytic philosophers like Alvin Goldman at Rutgers University. Only recently has mirror neuron work been undertaken that is relatively independent of the Parma group, with research into autism and psychopathy.[9]

The intertwining of these strands of mirror neuron research allows us to treat them as possibilities of a single research project. Analyzing that project as a whole and taking into account its history, which stretches from the 1980s to the present, makes visible the underlying structure of mirror neuron research, a structure that is obscured when analysis is restricted to only one of its component parts. As I will show, the organization of mirror neuron experiments made as conditions of research two forms of difference between the mirrored acts: *nonsimultaneity* and *incongruence*. References to these forms of differences have multiplied within the mirror neuron literature, and they have provided powerful resources for a number of important debates, for instance about the distinction between animals and humans in so-called higher order mirroring and the explanation of the origins of language. In this way a historical account of mirror neuron research that is attentive to its experimental conditions sheds new light on central features of the modern neural subject.

In the final part of the paper I will return to mirror neuron research on empathy and the emotional reading of mirror neuron function that it has legitimated. In that research the two forms of difference that I argue are essential to mirror neuron research—nonsimultaneity and incongruence—are not immediately visible. In emphasizing the immediate character of the mirroring response in empathy, mirror neuron researchers effaced the differences between the executed and the observed act. But as I will argue, when one looks at research into what were taken to be *pathologies* of empathy—psychopathy and autism—those differences resurface. The conceptual and experimental demands of research that sought to tease apart normal and pathological mirror neuron function required researchers to draw on the full complement of conceptual and experimental resources, exposing some that were otherwise hidden. Pathology is thus a privileged site for examining the underlying structure of scientific objects. I end the paper with a discussion of how this analysis recasts debates over the emotional or cognitive understanding of mirror neuron function.

## The Emergence of Mirror Neurons as a Research Program

THE PROCESS OF "mirroring" always involves difference. Even in the archetypal situation, my mirrored image confronts me as something foreign and inverted. But, depending on the type of mirroring, that difference can be expressed in a number of ways: Lacan's mirrors offer an "imaginary" wholeness, in contrast to the fragmentary self; fairground mirrors distort our shape and height. For mirror neuron theorists like Rizzolatti and colleagues, the differences that were constitutive of the neuronal "mirroring" they described were determined by the history and demands of their experiments. As I will show, those experiments required first the nonsimultaneity of observed and executed action, often produced by a process of inhibition, and second what I'd like to call a tolerance for incongruence.

To understand why these two forms of difference were central to mirror neurons, we need to take a brief detour through the broader tradition out of which mirror neuron research emerged: the mapping of the sensory-motor cortex. This history goes back to the work of Canadian neurosurgeon Wilder Penfield in the 1930s, 1940s, and 1950s. During that period Penfield and his coworkers mapped out the primary motor and somatosensory cortex, that is, the parts of the brain responsible for the processing of motor and tactile information. The maps were configured pictorially, as "homunculi," somatotopic maps with a positional arrangement of the body along the cortex (where the representation of the toes was adjacent to that of the foot, the lower leg, the upper leg, etc.). These images have since become iconic (fig. 9.1).[10]

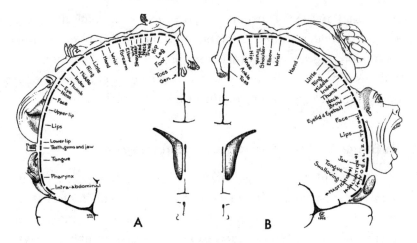

**Figure 9.1** Wilder Penfield's motor and sensory homunculi.
Source: Wilder Penfield and Theodore Rasmussen, *The Cerebral Cortex of Man* (New York: Macmillan, 1950), 44. Reproduced by permission of the Osler Library of the History of Medicine, McGill University.

Two assumptions of Penfield's maps are important for us here. First, the maps were either sensory or motor, not both.[11] The motor map indicated which part of the cortex controlled the movement of which body part; conversely the somatosensory map contained a tactile representation of the body. The separation was expressed spatially: the motor and somatosensory homunculi were located on the so-called precentral and postcentral sulci. Second, according to Penfield's model, the motor cortex coded only simple, elementary movements. The combination of such movements required the ordered excitation of multiple points on the motor map. For this it relied on a structure that lay *outside* the map, the higher order integrating system located between the sensory and motor cortices, a region sometimes referred to as the "associative areas."[12] In this view, then, with the associative areas at the center, the motor cortex was considered "peripheral and almost exclusively executive."[13]

In the 1980s Rizzolatti and his team in Parma worked to challenge the second assumption of Penfield's model; they disputed the attribution of such a lowly function to the premotor cortex.

Instead they thought that a part of the premotor cortex might be responsible for complex "actions," which Rizzolatti et al. defined "as a sequence of movements which, when executed, allows one to reach his goal."[14] In a pair of papers published in 1988 the researchers set out to demonstrate that "complex actions" were not dependent on an association system; they could be coded within the premotor cortex itself, even by a single neuron.

To provide evidence for their claims, Rizzolatti and colleagues worked with three macaque monkeys "selected for their docility" and trained to sit on a chair with their head fixed looking forward.[15] Electrodes were inserted into their brain, which recorded the activity of individual neurons during the experiment. The animals were surrounded by a "plexiglass perimeter" at arm's length. The plexiglass had nine holes arranged in three rows into which the experimenters could place pieces of food. The monkeys were then prompted to extend their arm to reach for the food, which counted for Rizzolatti as a complex "action" rather than a simple "movement." Through the varying locations in space of the objects in the nine perimeter holes, and an additional set of variations of the test (e.g., presenting food to the animal closer to the body, and with different experimental apparatuses), researchers were able to relate a large set of actions to the activity of individual neurons.[16] Based on these data, they mapped out the function of the premotor cortex.

We can see how the maps Rizzolatti produced could serve as evidence for the complexity of the premotor cortex. The maps showed two things. First, the premotor cortex, more specifically the rostral and caudal areas of its inferior part (areas F4 and F5), had its own somatotopy, which lent a certain independence to it.[17] Second, as the researchers pointed out, in the premotor areas "neuron activity [was] frequently related to movements involving more than one articulation, [and] the maps [were] expressed in terms of body parts controlled by the neuron encountered in a given penetration."[18] The maps were thus representations of "active movements" or, in the later terminology, "actions" (fig. 9.2).[19]

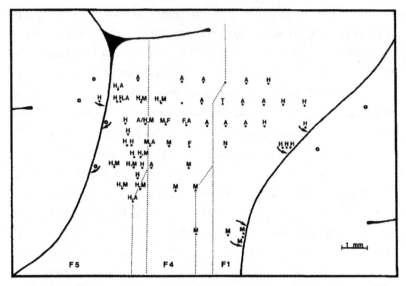

**Figure 9.2**    Brain map of areas F1, F4, and F 5. The letters indicate parts of the body: A = arm; F = face; H = hand; M = mouth; N = neck; T = upper trunk. Source: M. Gentilucci, L. Fogassi, G. Luppino, M. Matelli, R. Camarda, and G. Rizzolatti, "Functional Organization of Inferior Area 6 in the Macaque Monkey. I. Somatotopy and the Control of Proximal Movements," *Experimental Brain Research* 71 (1988): 478. Reprinted with kind permission from Springer Science and Business Media.

Rizzolatti's challenge to the second assumption of Penfield's motor maps (that they did not code "actions") also led him to challenge the first: the clean distinction between sensory and motor areas. The actions Rizzolatti elicited from the macaques had an inherent sensory element. They involved both the visual recognition of the stimulus in space (presented in one of the perimeter holes) and the proprioceptive experience of grasping the object.[20] In the 1988 papers the Parma team described a similar mixture of sensory and motor elements at the level of the neurons.[21] The majority of neurons in area F4 was "strongly responsive to tactile stimuli" and "related to proximal and facial movements."[22] Moreover the stimuli and actions were closely related: for instance, the neurons that controlled proximal movement (i.e., movement within the space close to the body) were "triggered by stimuli presented in the animal's peripersonal space." "Thus, an object presented in a particular spatial position activate[d] the neurons controlling the motor act 'reach' and, if the motivation [was] sufficient, these neurons [would] bring the arm in the space position where the stimulus [was] located."[23] The researchers concluded that for F4 neurons "their input-output relationship [was] very complex." The difference between Penfield's traditional "executive" primary motor cortex and the premotor cortex was thus "quite clear."[24]

## The Nonsimultaneity of Observation and Action in the Mirror Box

THE FUNCTIONAL COMPLEXITY of the reaching-grasping neurons explains the introduction of the first form of difference into our story: nonsimultaneity. In 1990 the Parma team turned their attention to the anterior superior part of the premotor cortex (area F7).[25] As before, these neurons seemed to be closely related to reaching and grasping movements, that is, complex actions, which were tested with an experimental apparatus (the perimeter test) similar to that in the 1988 papers. In the 1990 paper, however, the researchers adopted more rigorous means to prove that the neurons were functionally complex; they had to discount the possibility that the neurons' activities could be explained in sensory or motor terms alone. This required two conditions: a recording of neural activity clearly attributable to the sensory stimulus and a recording of activity attributable to the movement. And because these two conditions were mutually exclusive, they had to be separated temporally.

The researchers related the neuronal activity to three moments in the experiment: the initial awareness of the sensory stimulus (marked by a saccade, which is an eye movement), the beginning of the reaching-grasping movement, and the end of the movement. In figure 9.3 these three moments are represented by a triangle, a dotted vertical line, and a heavy vertical line, respectively. The three moments delimited two distinct phases of neuronal activity: stimulus without movement, and movement.

As figure 9.3 shows, the researchers noted a change in neuronal activity—represented by white space[26]—after the presentation of the stimulus. Importantly the activity of the neuron was similar during both the premovement phase (from the initial awareness of the stimulus to the beginning of the reaching movement, triangle to dotted line) and during the movement (dotted line to heavy line).[27] Rizzolatti concluded that the neurons in area F7 were functionally complex.

Even given this temporal separation, Rizzolatti wasn't entirely satisfied that he had identified both sensory and motor properties. Because the neurons were located in the premotor cortex, he was most concerned to discount the possibility that the neurons' activity during the premovement phase could be attributed to motor activity.[28] One such concern revolved around "set-related neurons." Set-related neurons had been found in various motor areas and showed activity before the movement began. But this premovement activity "reflect[ed] the motor aspect of the movement preparation rather than sensory or motivational factors."[29] If the reaching-grasping neurons were like set-related neurons, the premovement activity wouldn't indicate that the neurons also had sensory properties. I will not go into the details of how Rizzolatti and his team sought to distinguish the two sensory and motor responses; for my argument it is sufficient to know that they took great care to do so: they needed a clear and sharp distinction between sensory and motor phases.[30]

In the 1990 paper the nonsimultaneity of stimulus presentation and movement, and thus the delimiting of two distinct phases of neuronal activity, was determined by observation. In a paper published two years later that nonsimultaneity was produced by the experimental apparatus. In "Understanding Motor Events: A Neurophysiological Study," the paper that first described the characteristics of mirror neurons (without yet using their name), Rizzolatti and his coworkers

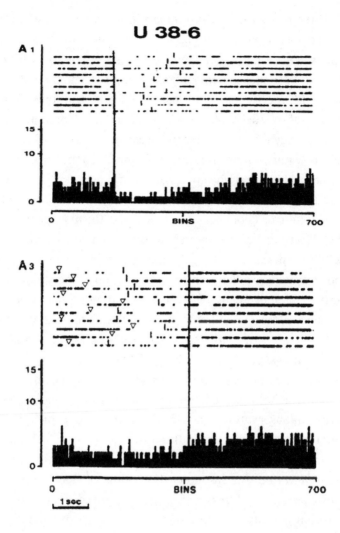

**Figure 9.3** Neural firing is inhibited before and during a reaching-grasping movement. Every row in this plot is a single trial. Every black dot is the occurrence of an action potential within the trial. Ten individual trials are shown. The triangles signify the saccadic eye movement triggered by stimulus presentation; the dotted line designates the beginning of movement; and the solid line designates the end of the movement.

Source: Modified from G. Rizzolatti, M. Gentilucci, R. M. Camarda, V. Gallese, G. Luppino, M. Matelli, and L. Fogassi, "Neurons Related to Reaching-Grasping Arm Movements in the Rostral Part of Area 6 (Area 6aβ)," *Experimental Brain Research* 82 (1990): 341. Reprinted with kind permission from Springer Science and Business Media.

repeated their experiment with a new setup. Now food was placed under a geometric object in a box. The front door of the box consisted of a one-way mirror. When the monkey pressed a switch, the inside of the box was lit so that the mirror became transparent, presenting the monkey with the sensory stimulus. Provided he continued to press the switch, after a delay of 1.2 to 1.5 seconds the door would open, and the monkey could reach for the object and thus access the food. The one-way mirror thus experimentally produced the three dividing lines that had been described in the earlier paper: the presentation of the stimulus (mirror-door becomes transparent), the beginning of the movement (mirror-door opens), and the end of the movement (grasping the object). But unlike the relatively messy and varying timings of the 1990 experiment, where the triangles didn't line up and there was no consistent gap between the presentation of the stimulus and the start of the movement, here it could be more closely controlled: the two phases, "seeing without moving" (glass door shut) and "moving" (glass door open), were neatly separated.

In the analysis and comparison of these two phases we can already see the sensory-motor basis that would become central for the mirroring neurons: neurons that both responded to sensory stimuli (observation condition) and were active during motor action (execution condition). But the structure of the experiment makes clear why the presentation of the stimulus and the action couldn't be simultaneous. The evidence that the neurons had both sensory and motor properties required the comparison of neuronal activity during two mutually exclusive conditions: stimulus without movement, and movement. This reliance on the nonsimultaneity of the stimulus and the executed action would carry over into the research on mirror neurons.

## Incongruence

AT THIS STAGE we are still an important step away from mirror neurons. The two experimental conditions during which neuronal activity was recorded and compared remained starkly different: On the one hand the researchers measured the activity of the neuron during a complex action; on the other they measured it during the presentation of a simple sensory stimulus, an object placed in a particular location in the monkey's visual field. Given the disparity between what was observed and what was executed, it was difficult to make the claim that there was any mirroring between them. The emergence of mirroring as an explicit theme would have to wait for a chance event.

In running the experiment the researchers periodically had to enter the scene to pick up the food or place it inside the mirror box. During this process (officially outside the limits of the experiment) the monkeys remained in place and the recording equipment was on. The electrodes thus caught the activity of the monkey's F5 neurons in response to the researchers' actions. As the 1992 paper noted, "in the absence of any overt movement of the monkey," the behaving scientist activated a "relatively large proportion of F5 neurons," which had been active when the monkey performed the same movements.[31] That is, the researchers had inadvertently substituted a new stimulus, and—this was the important change—the observed action (researcher reaching for the object) was now similar to—in Rizzolatti's language, "congruent" with—the monkey's (monkey reaching for the object); they mirrored each other.[32] Following this discovery the researchers began to "perform a series of motor actions in front of the animal."[33]

The increased similarity between observed and executed actions made the "congruence" between them a central concept in the research. Already in the 1996 paper Rizzolatti distinguished between two types of mirror neurons based on their congruence.[34] In the first class, the *broadly congruent* (the most common type, accounting for over 60 percent of mirror neurons), neurons could be activated by a varied range of actions, for example reaching, grasping, and rotation of the hand; in the researchers' words, "There was a link, but not identity, between the effective observed and executed action." In the second class, the *strictly congruent*, the observed movement and the executed movement correlating with the neuronal activity "corresponded both in terms of general action (e.g. grasping) and in terms of the way in which the action was executed (e.g. precision grip)."[35]

The responses of a strictly congruent neuron are presented in figure 9.4. The neuron was tested under three conditions: in A the experimenter placed a raisin on a tray; in B he grasped the raisin; and in C the monkey performed the same grasping action. Actions B and C were "congruent," and in both cases the neuron showed a "response inhibition of the spontaneous discharge" (which can be observed by the absence of activity immediately after both observing, as in B, and executing, as in C, the grasping).[36] The action A was not, however, sufficiently similar to C, and in that case the neuron showed no change in activity. For "strictly congruent" neurons the observed actions had to reach a threshold similarity to the executed action in order for the neuron to fire.

The term *congruent* seems appropriate to this situation. In geometry two figures are congruent when they have the same shape and size but different locations and orientations. So too in the mirror neuron experiment the same action (grasping) was performed by two different subjects, in this case the macaque and the experimenter. But because the actions were performed by different subjects, even at the level of the action there could be no identity. In fact even strictly congruent mirror neurons needed to respond to a range of actions that were only broadly similar to the executed action coded by the neuron; they had to have a tolerance for incongruence. This tolerance for incongruence was crucial both for the research trajectory—after all, when they first noted neuronal response to the observation of a "congruent" movement, they were actually testing for its response to an entirely incongruent static stimulus—and for the experiment

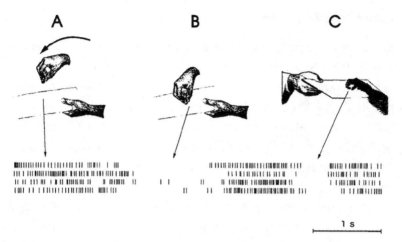

**Figure 9.4**   Example of a highly congruent mirror neuron.
Source: G. Rizzolatti, L. Fadiga, V. Gallese, and L. Fogassi, "Premotor
Cortex and the Recognition of Motor Actions," *Cognitive Brain Research* 3
(1996): 135. Reprinted with permission from Elsevier.

itself, where the experimenters could only hope to activate the
neurons if they didn't have to produce a direct copy of the animal's action.

The nonsimultaneity and incongruence of the observed and
executed actions are hardly surprising given the nature of the
experiment. It is impossible to imagine that the two actions
could be absolutely simultaneous or identical. But as I have
suggested, these forms of difference were not simply noise in
the experiment, something to be reduced as much as possible.
Without the nonsimultaneity of the two actions it would not
have been possible to attribute sensory properties to the neuron. It was not that the two acts *were* not simultaneous; they
*could not* have been. And without a tolerance of incongruence,
the researchers would not have been able to experiment on and
thus analyze the neurons' sensory properties. The two forms of
difference were conditions of the discovery of and experimenting on mirror neurons.

## Human-Animal Distinction

NOT ONLY DID nonsimultaneity and incongruence play a central role in the discovery of mirror neurons; they also provided conceptual resources for some of the most important debates within the field. Not least, despite the robust and multiple connections between animals and humans that mirror neuron research has brought to light, these two forms of difference have opened space for others to argue that mirror neurons hold the key to human specificity.

The history of mirror neuron research, as I have been telling it here, certainly tends to challenge the idea that mirror neurons set humans apart from animals, not least because they were first discovered in monkeys. The relationship becomes even more complicated when we remember that the first mirroring took place between humans and animals, when the recording device picked up neuronal activity in the macaque brain as the scientists set up the experimental system.[37] Moreover, in the paper that first suggested the existence of mirror neurons in humans, a collaboration between the Parma group and researchers at the University of Milan, the researchers relied on homologies between monkeys and humans to make their case.[38] Given this closely entwined nature of human and animal research, how have scientists justified their claims that mirror neurons set humans apart? The answer lies in a judicious use of the two forms of difference I discussed earlier: nonsimultaneity and incongruence.

## Higher Order Mirroring and Nonsimultaneity

THE MOST PROMINENT scientist to claim that mirror neurons provide the key to our humanity is Ramachandran, one of the few who contributed to the debate about mirror neurons without sustained collaboration with the Parma center. In 2001 Ramachandran discussed the "great leap" (sometimes called the "big bang") of human evolution, which was a sudden rise in technological development and cultural expressions such as cave art, clothes, and new kinds of dwellings about forty thousand years ago.[39] He asked why this happened only then, rather than 250,000 years ago, when the hominid brain had reached its present size. The common answer was that, due to a sudden genetic shift, previously unconnected functional areas of the brain were suddenly able to work together. Ramachandran suggested a modification of this view: if there was a genetic change, it concerned the mirror neuron system, increasing its "sophistication" and therefore its "learnability."[40]

It is not exactly clear what "sophistication" means here, but from what Ramachandran argued, he seemed to be referring to the *human ability to imitate highly complex actions.* As he wrote, these "sophisticated" mirror neurons, and the advanced imitation capacities they enabled, allowed developments such as "tool use, art, math, and even aspects of language" to "spread very quickly through the population."[41] The emphasis here was not so much on invention as dissemination. Ramachandran admitted that the same or similar innovations might as well have occurred with "earlier hominids" like *Homo erectus* or Neanderthals, but, he suggested, because these hominids possessed a less powerful mirror neuron system, the innovations "quickly drop[ped] out of the 'meme pool.'"[42] The distinction between humans and animals, and between *Homo sapiens* and other members of the same genus, then, seemed to be located in their ability to match ever more closely the observed action.

As the discussion developed, and the analysis of higher order mirroring was picked up by Iacoboni at the UCLA Brain Mapping Center, it became clear that the key to this process was inhibition (separating action-execution and action-observation temporally). Drawing on the work of Dutch social psychologist Ap Dijksterhuis, Iacoboni distinguished two forms of imitation in humans: "low road" imitation, a straightforward form of imitation like grasping a cup; and "high road" imitation, a form of complex automatic mimicry where subtle forms of imitative behavior (e.g., set off by thinking of a college professor on the one hand, or soccer hooligans on the other hand) had an effect on cognitive performance.[43] Iacoboni didn't believe that these higher forms of imitative behavior could be explained by the class of mirror neurons that the Parma team had discovered in monkeys. Rather, he suggested, an additional element was needed to regulate the simpler mirror neurons. He called these higher order neurons "super mirror neurons," to indicate that they worked as a "functional layer 'on top of' the classical mirror neurons, controlling and modulating their activity."[44]

The existence of such "super mirror neurons" was suggested to Iacoboni by the results of a collaborative study with the electrophysiologist Itzhak Fried at Tel Aviv University.[45] The paper "Single Neuron Responses in Humans during Execution and Observation of Actions" that resulted from this collaboration is considered a milestone in the history of mirror neuron research, for it provided the first direct evidence for mirror neurons in humans.[46] But what interests me about the paper here is that the researchers found evidence for super mirror neurons. They took advantage of a group of patients suffering from drug-resistant frontal lobe epilepsy (their epileptic foci were located in the medial frontal and temporal cortices). The last resort treatment of these patients was surgery, and in order to locate the exact focus of the epileptic activity and tell the surgeon where to cut, electrodes were routinely implanted in various brain regions so that brain activity could be measured over an extended period of time. The situation provided an unrivaled opportunity for studying areas of the brain with far greater precision than was usually allowed; Iacoboni could benefit from the "exquisite resolution of single cells" recorded from these patients while they completed various tasks.[47]

In the study, a collaboration between researchers at UCLA, UC Davis, and Tel Aviv University, subjects were asked to observe various grasping actions and facial gestures on a computer screen (observation conditions); in addition they were cued to perform the same actions in response to a visually presented word (execution conditions). The scientists identified a set of mirror neurons in the medial frontal and temporal cortices, which responded during both the execution and the observation of the same action. These, however, could be divided into two groups, one that had similar responses in both action-execution and action-observation and another where the cells responded "with excitation during action-execution and *inhibition* during action-observation" (fig. 9.5).[48] It was these latter cells that Iacoboni would later call "super mirror neurons," which the researchers suggested might play a role in controlling "unwanted imitation."[49] As they wrote, "Mirroring activity, by definition, generalizes across agency and matches executed actions performed by self with perceived action performed by others. Although this may facilitate imitative learning, it may also induce unwanted imitation. Thus, it seems necessary to implement neuronal mechanisms of control. The subset of mirror neurons responding with *opposite* patterns of excitation and inhibition during action-execution and action-observation seem ideally suited for this control function."[50]

The point is that Iacoboni explained the ability of super mirror neurons to "organiz[e] simpler imitative actions—the low road—into complex forms of imitative behavior—the high road" by their ability to separate the observed and the executed act.[51] But of course this separation was not a new addition to mirror neuron research. In breaking the immediacy of the relationship between observed and executed acts in super mirror neurons, Iacoboni was only maximizing a similar mediation at the heart of all mirror neurons (including those in macaques): the action inhibition during observation and thus nonsimultaneity between observation and execution that had been central to their discovery in the first place.

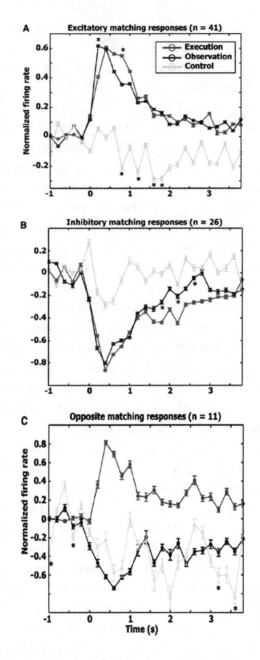

**Figure 9.5** "Super mirror neurons" showing inverse mirroring.
Source: R. Mukamel, A. Ekstrom, J. Kaplan, M. Iacoboni, and I. Fried, "Single-Neuron Responses in Humans during Execution and Observation of Actions," *Current Biology* 20 (2010): 754. Reprinted with permission from Elsevier.

## Language and Incongruence

SIMILARLY THE TOLERATION of incongruence has also played an important role in the human-animal debate, especially as relates to the origin of language. The origin of language is not a widely studied question in mirror neuron research—we are essentially dealing with a handful of papers written in the 1990s of which a short speculative paper by Rizzolatti and the computer scientist Arbib from 1998 is the most important one—but the arguments used are nonetheless instructive.[52]

As mentioned in my earlier discussion of the 1996 paper, the Italians had argued for the existence of mirror neurons in humans by drawing on homologies with monkey brains. Significantly F5 in monkeys, the area that contained mirror neurons, corresponded to area 44 in humans, also called Broca's area. Since Broca's area has long been regarded as the area of the brain related to motor speech, the homology seemed to suggest that mirror neurons might play a role in the emergence of language. To understand this connection, Rizzolatti and Arbib engaged with the idea that there was a sharp divide between human and animal speech.[53] Vocalization in nonhuman primates was mediated by the cingulate cortex and the diencephalic and brain stem structure, located deep within the brain, far from Broca's area. In addition animal vocal calls seemed to serve a different form of communication, which aimed not at a particular individual, like speech, but at a group at large. Finally, monkey vocalization was related to instinctive and emotional behavior, whereas speech was not.[54] While agreeing with the broader argument, Rizzolatti and Arbib suggested another evolutionary pathway: human speech evolved out of animal gestures, gestures controlled by mirror neurons.[55] The mirror neurons in motor area F5 might then be the "neural prerequisite for the development of inter-individual communication and finally of speech."[56] This was the significance of the supposed homology between F5 and Broca's area.

Rizzolatti and Arbib argued that mirror neurons could explain primary communicative gestures, but to do so they had to rely on the incongruence between the mirrored actions. When confronted with an action "of particular interest," the mirror neuron system would "allow a brief prefix of the movement to be exhibited." Thus in seeing ingestive processes, the monkey would perform "lipsmacks" or "tonguesmacks." In similar ways "a primitive vocabulary of meaningful sounds could start to develop."[57] Later experiments suggested a similar process in humans. For example, Gentilucci et al. presented participants with two 3-D objects, one large, one small, and asked them to open their mouth when seeing them. They found that lip aperture increased when the movement was toward a large object and decreased when directed at a small object.[58] What is important for my argument here is the way in which differentiation or the lack of absolute congruence was central to the process; the mirrored "prefix" (lipsmacks) was only weakly congruent with the act observed (eating). Such differentiation allowed the connection and yet difference between the signified object and the signifying sign, and by extension the development of a linguistic system where even this tenuous relationship of resemblance was given up. That is, the process of multiple incongruent mirrorings eventually allowed a purely conventional relationship between signifier and signified. In Rizzolatti and Arbib's words, "Sounds acquired a descriptive value."[59]

### Empathy and the Effacing of Difference

WHILE TEMPORAL DISPLACEMENT and incongruence, forms of difference fixed by the initial experimental setup of mirror neuron research, have played important roles in debates around human-animal difference, they have been far less visible in the most prominent strand of mirror neuron research: on empathy.[60] In fact, despite researchers' claims, empathy research required a significant refiguring of what mirroring involved. The best-known paper connecting a mirror neuron mechanism to empathy (a "mirroring" of other people's emotions) was published in 2003 by Marseille-based neuroscientist Bruno Wicker in a collaboration with Rizzolatti, Gallese, and Christian Keysers at Groningen, Netherlands: "Both of Us Disgusted in *My* Insula." Running separate fMRI trials on human subjects for the observation of the facial expression of disgust on the one hand, and the actual experience of disgust from exposure to "stinking balls" on the other hand, the researchers found that some regions of the anterior insula were activated in both cases. The researchers asserted that the same principles underlying the function of "classical" visuomotor mirror neurons in F5 applied to those responsible for these emotions: "As observing hand action activates the observer's motor representation of that action, observing an emotion activates the neural representation of that emotion." The understanding of emotions took place by "matching felt and observed emotions" (fig. 9.6).[61]

At first sight the experiment looks similar to the earlier ones: a comparison of neuronal activity during observation and execution. But the French-led team diverged from previous papers in a crucial way. As we saw in the earlier experiments involving actions, it was a condition of the analysis that during observation of an action (the experimenter grasping a raisin), the subject's own performance of that action (the macaque grasping the raisin) had to be delayed, otherwise it would have been impossible to exclude the possibility that the neuronal activity was correlated with the executed, not the observed, action. This was what I called the nonsimultaneity condition. Transposed to the empathy experiments, this would suggest that to be able to prove the existence of mirroring, the researchers would have to measure neuronal activity in response to the observed emotion during a period when the subject wasn't feeling (executing) that emotion herself.

Wicker et al.'s experiments, however, did not recognize this condition. This can be seen in a slippage in their terminology. For the

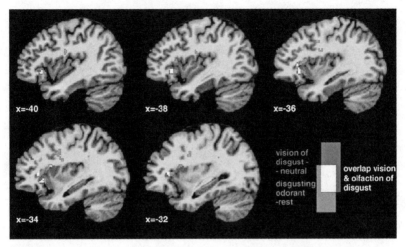

**Figure 9.6** Sagittal sections through the averaged left hemisphere of the fourteen participating subjects in the study. The white patches indicate the overlap of the vision and olfaction of disgust, which are located in the insula. *Source:* B. Wicker, C. Keysers, J. Plailly, J.-P. Royet, V. Gallese, and G. Rizzolatti, "Both of Us Disgusted in *My* Insula: The Common Neural Basis of Seeing and Feeling Disgust," *Neuron* 40 (2003): 660. Reprinted with permission from Elsevier.

majority of the paper the authors are clear that observation merely activates the "neural representation" of an emotion, and this is what allows the subject to "understand" it. But at a number of crucial moments, especially when working out the larger implications of the research, they suggest that this was more than just a "representation." During the "observation condition," when the subject was presented with a video clip of a disgusted face, the researchers claimed that the subject also felt disgusted, so the emotion was "shared." The activation of the emotion (its feeling), not just its neural representation, is what allowed Wicker et al. to endorse the "hot hypothesis" of emotion understanding. When an observer saw a "conspecific" looking disgusted after the consumption of food, the observer inferred "automatically" that the food should be avoided because they *experienced* the disgust at the same time.[62]

The slippage between the "neural representation" of an emotion and that emotion itself also directly concerns the second form of difference: incongruence. When Wicker et al. implied that the neuronal representation *was* the emotion, they shut down the possibility that the neuronal representation could correspond to a range of different feelings. It is telling that in the paper the authors did not actually ask the subjects about their emotional response to sniffing the odorants or observing the actors' facial expressions; the fMRI results were sufficient evidence of the emotion.[63] Ruth Leys has made a parallel criticism in her questioning of the cogency of Wicker et al.'s experiment (together with the follow-up study by Jabbi et al.).[64] She explains the failure of mirror neuron researchers to ask their subjects about their emotional states by their assumption that there was a direct and reliable relationship between so-called basic emotions and facial expressions. According to this assumption, Leys argued, the researchers collapsed emotions into corporeal states and neglected the "social-transactional character" of emotional (re)actions.[65]

If in much empathy research we see a downplaying of the constitutive differences of the mirror neuron paradigm, that does not mean that they were fully effaced there. Indeed we can see a return to the themes of nonsimultaneity (through inhibition) and incongruence at the margins, most importantly in mirror neuron researchers' attempts to understand disturbances of empathy: psychopathy and autism. As is often the case in medicine, researchers had to draw on the full arsenal of conceptual resources available to them in order to understand the complexity of pathology and its relationship to normality. For this reason the study of research into psychopathy and autism provides privileged access into the structure of mirror neuron research into empathy more generally.[66]

## Psychopathy

IN RECENT YEARS a number of smaller research groups have begun to investigate the relationship between mirror neuron systems and psychopathy; among these is a collaboration between the University of Montreal and Harvard Medical School led by Alvaro Pascual-Leone and a research group at University of Groningen in Holland led by Christian Keysers.[67] The use of mirror neurons, more specifically the ability to produce nonsimultaneity through inhibition, helped explain the dual nature of psychopathy: psychopaths seemed at the same time acutely aware of the emotions of others (which allowed them to be so charming) and profoundly indifferent to them (which accounted for their ability to harm others). In a more popular presentation of his work, Keysers used the fictional Hannibal Lecter from *The Silence of the Lambs* to describe the condition: "Despite the varnish of sophistication . . . Lecter is capable of horrible crimes." What makes psychopathy so disturbing to us is not simply the discovery that psychopathic individuals lack empathy; it is also the insight that they "combine a talent for manipulation with a lack of remorse."[68]

Keysers's graduate student Harma Meffert and their colleagues demonstrated this dual character of psychopathy and its relationship to the mirror neuron system in 2013 in an fMRI study on convicted psychopathic offenders in Holland.[69] In the "observation condition," the subjects were asked to watch videos of two hands interacting with each other in different ways, representing "love" (hands caressing each other), "pain" (one hand hitting the other), and "social exclusion" (one hand pushing away the other hand), while their brains were scanned. In the "experience condition," subjects experienced matched conditions on their own hands, being alternately caressed, hit, or pushed away by the experimenter. During all this the activity of their brain was recorded.

As the results suggested, the psychopathic group showed reduced activity in regions involved in experiencing emotions, the supposed mirror neuron areas, what the researchers termed reduced "vicarious activity."[70] The psychopaths did not "feel" the emotions of the people they observed. But the study also showed that this didn't have to be the case, and this allowed the researchers to draw conclusions about the seductive-merciless character of psychopaths. In a third condition (the "empathy condition"), subjects were explicitly instructed to empathize, to "feel with the receiving . . . or the approaching . . . hand" on the videos. The researchers found that these "explicit instructions to empathize significantly reduced the group differences with regions associated with vicarious activations."[71] The mirror neuron system in psychopaths, which was normally "switched off," could also be switched on.[72] In fact this possibility held the key to the psychopaths' ability to be charming as well as cruel.[73]

This confirmed a similar observation that had been made five years earlier in normal individuals. In their 2008 study the Harvard-Montreal group of Pascual-Leone (a leader in the development of transcranial magnetic stimulation, or TMS) recorded TMS-induced motor evoked potentials (MEP) in a group of male college students.[74] Based on results from previous studies, they assumed that the observation of pain in others led to a decreased MEP amplitude in normal subjects, which was considered a measure of mirror neuron activity. To determine character traits all subjects also took the Psychopathic Personality Inventory, which situated the subjects along several axes measuring their "Machiavellian egocentricity," "coldheartedness," and "stress immunity." Somewhat surprisingly the team found that individuals who scored highest on the coldheartedness scale displayed the largest modulation of cortical excitability.[75]

To explain their results the Montreal-Harvard group appealed to a distinction made by A. Aventani et al. in 2005 between "sensory empathy" and "emotional, state or trait empathy."[76] "Sensory empathy" was the "strict ability to understand the affective, sensory or emotional state of another individual"; "trait empathy" made that information "available to the observer for an emotional/affective response." They suggested that their conclusions—using a distinction between understanding and feeling of emotion—were consistent with data by R. Blair, who in 2005 had suggested that in psychopathy "motor empathy" and "theory of mind" were unaffected, whereas "emotional empathy" was impaired (here "motor empathy" roughly mapped onto "sensory empathy" in Aventani's distinction). The high modulation of cortical excitability among the "cold-hearted" subjects suggested that they were sensorially empathetic, which enabled their "notorious manipulative nature" and "their ability to exploit weaknesses in others." In contrast "trait empathy" seemed to be "maladaptive" in psychopaths.[77] The subjects' pathology could thus be traced to their ability to distance themselves emotionally from the actions of others, to separate the knowledge of the observed act from the feelings they had themselves. The nonsimultaneity of observed and felt emotion in "sensory" empathy thus became key to understanding psychopathy.

## Autism

WHILE PSYCHOPATHY RESEARCH suggested the existence of a form of empathy and mirroring that involved separation of observed and felt emotions, a parallel strand of research suggested that our two forms of difference—nonsimultaneity and incongruence of execution and observation—were not merely pathological but were necessary for normal mirroring. Deeply connected in this tradition to the question of empathy, autism has a similarly prominent profile in the research.

Experimental evidence for a connection between mirror neurons and autism came from three research groups in the mid-2000s. In 2004 a group of researchers around neuroscientist Riitta Hari at the Helsinki Institute of Technology noted that in a magnetoencephalography study individuals with Asperger's syndrome (a condition related to autism) manifested delayed activation of the mirror neuron system in the inferior frontal lobe during an imitation task.[78] A year later, using a method involving "mu oscillations" to determine mirror neuron activity, Ramachandran and his colleagues at UC San Diego compared normal and autistic patients while opening and closing their own hand at a specific rate and observing the same action on a video screen.[79] The lack of "mu suppression" in the autistic patients during observation suggested to the researchers a "possible dysfunction in the mirror neuron system," what Ramachandran later presented as the "broken mirror" hypothesis.[80] Finally, in 2006 Justin Williams at the Department of Child Health at the University of Aberdeen and his coworkers (among them the evolutionary psychologist Andrew Whiten and David I. Perret, an expert in monkey neurophysiology at St. Andrews) used fMRI to provide evidence for a link between mirror neurons and autism. Building on a protocol used by Iacoboni et al. in 1999 to identify the neural substrate of imitation and mirror neuron function, they found a different pattern of brain activity in autistic individuals during an imitation task.[81]

The authors of these three studies discussed in considerable detail the neurophysiological changes seen in the disturbed mirror neuron system. But on the specifics of the failures (behavioral, cognitive, emotional, etc.) they were less clear. It remained undetermined how the symptoms of autism could be explained by a failure of mirroring. What did it mean to say that autism was caused by "broken mirrors"? For the most part Ramachandran and Williams outsourced this part of the project, relying on other studies to show the breakdown of mirroring in autistic patients. Claiming that mirror neuron impairment helped explain the full range of symptoms observed in autism, including imitation, language, theory of mind, and empathy, Ramachandran cited a range of works for each field.[82] Similarly Williams referred back to one of his earlier articles on the topic, which contained a brief survey of autism research.[83]

This literature review (Williams et al. 2001) discusses the complex relationship between autism and mirroring. Williams et al. referred to studies showing that "people with autism do not readily imitate the actions of others," especially when those actions were "complex." But while this problem fit with the idea of an absence of mirroring in autism, the researchers also made reference to another type of pathology, referring to studies that described how autistic patients showed "inflexible and stereotyped behavior and language," such as "copying actions," "obsessive desire for sameness," and "more stereotyped mimicking, such as echolalia." For example, autistic patients might only repeat a question instead of answering it. Here the excess of mirroring, not its lack, was the problem. Williams et al.'s explanation for this second type of pathology should hold our interest, because it recalls our previous discussion of super mirror neurons. They argued that in such cases the mirror neuron system "might be evidencing poor modulation"; that is, the "inhibitory components" of the system were damaged.[84] The implication was that healthy mirroring required inhibition, a separation or nonsimultaneity between observed and executed action.

We see a reference to the second form of difference, incongruence, in a later paper by Ramachandran and his colleagues at the UCSD Center for Brain and Cognition. In an editorial for the journal *Medical Hypotheses* published in 2007, they emphasized the ways mirror neurons allowed the subject to discover commonality across different realms. They then extended the argument to help explain the understanding of metaphors. Quoting Shakespeare's line "Juliet is the sun," they claimed that we could understand it because we can "reveal the common denominator of radiance and warmth."[85] In autism this ability fails; individuals with autism were often unable to understand metaphors and interpreted them literally.[86] As before, we see indeterminacy with respect to the failure of mirroring. Did the autistic patients lack functioning mirror neurons? Or was the problem that those mirror neurons were too congruent and were activated only when the similarity between the two elements reached a certain threshold? Also as before, the implication was that normality was marked by a certain tolerance for incongruence between the observed and the executed conditions, in this case the representation of the sun and the feeling of warmth in love. The two couldn't be immediately related.

Autism thus seemed to be the inverse of psychopathy: the former at times promoted an excessive identification between the observed and executed acts; the latter made the separation too absolute. Researchers emphasized this relationship. As Pascual-Leone's group pointed out in their 2008 paper, whereas in psychopathy emotional empathy was impaired with theory of mind and motor empathy intact, in autism emotional empathy was spared, and theory of mind and motor empathy showed deficits.[87] The two conditions were mirror images of each other.[88] In both cases, however, nonsimultaneity and incongruence were the guiding principles. The analysis of autism suggested that normality might require a form of distancing and incongruence; the study of psychopathy suggested that distancing should not be pushed too far. In both cases the diseases were understood through a refiguring of the two forms of difference that were constitutive of the mirror neuron project.

## Emotion and Cognition

NOT ONLY DID different mirror neuron researchers adopt different approaches to the questions of nonsimultaneity and incongruence; their attitudes tended to correlate with the stance on the emotion/cognition question. Wicker and his colleagues, who downplayed these differences, used their analysis of empathy to suggest that humans had a "hot," noncognitive, understanding of others. The emotional and noncognitive nature of the process made it a likely candidate for the "evolutionary [sic] oldest form of emotion understanding." It was a "primitive mechanism," shared by monkeys and humans to protect them from food poisoning.[89] Wicker et al.'s argument was part of a broader shift in neuroscience and psychology, in which researchers asserted the priority of direct and automatic processing of information.[90] Take, for instance, the 1998 paper by Gallese and Alvin Goldman, a philosopher at Rutgers University, who sought to explain how "mind reading" was possible in humans. Gallese and Goldman favored the "simulation theory" (ST) over the competing "theory theory" (TT).[91] According to theory theory, "mental states [were] represented as inferred posits of a naïve theory."[92] Through a process of logical inference from a person's actions and appearance, those actions could be understood and future behavior predicted. Here inference and cognition took center stage. By contrast simulation theory suggested that we understood the actions of others through direct internal "representation" of the other's mental state, with no theorizing involved. We come to read other minds by "putting [ourselves] in the other's shoes."[93] It is easy to see how the mirror neuron system fit this latter model. Execution-observation matching suggested that observing an action was neurologically similar to executing it, and by extension in seeing the expression of an emotion we "felt" it too. Gallese was explicit in a paper published in 2001. To empathize, in his account, was essentially to understand: "My proposal is that also sensations, pains and emotions displayed by others can be *empathized, and therefore understood,* through a mirror matching mechanism."[94]

For those researchers who emphasized nonsimultaneity and incongruence, however, mirroring was often presented as a nonemotional, cognitive process. Take the Harvard-Montreal researchers on psychopathy: they opposed "sensory empathy" to "emotional empathy" because the former involved inhibition. In a similar way the LA researchers linked inhibition to higher order imitation (think college professor). The appeal to incongruence too tended to accompany a de-emphasis of the emotional. It was related to the emergence of language in Rizzolatti's account and to the understanding of metaphor in Ramachandran's.

Mirror neuron research has relied on a complex articulation of similarity and difference. Though, as we have seen, there was a tendency to relate mirroring to repetition, emphasizing its immediate emotional, noncognitive nature, at the heart of the mirroring process we can also discern the negotiation with difference. This difference was inscribed in the very experimental setup in which mirror neurons had been discovered and was constituted by an explicit separation and nonsimilarity between the mirrored actions. A consideration of this differentiation helped distinguish humans from other animals and explain a number of intrahuman differences. These elements may have been downplayed in the empathy research, which has dominated the public understanding of mirror neurons, but this does not mean that they were entirely absent there. For though "mirroring" can be understood as unthinking imitation, we should remember that thought is its own form of reflection.

# Endnotes

I would like to thank Edward Baring, Michael Gordin, Leor Katz, Ruth Leys, and the two editors for their insightful comments and suggestions on this chapter.

[1] The class of premotor neurons that showed the mirroring phenomenon was first described in G. DiPellegrino, L. Fadiga, L. Fogassi, V. Gallese, and G. Rizzolatti, "Understanding Motor Events: A Neurophysiological Study," *Experimental Brain Research* 91 (1992): 176–80.

[2] Area 6 is another name for the premotor cortex. It is divided into four subsegments, the rostro-ventral part (F5), the caudal-ventral part (F4), and two dorsal parts (F7 and F2). In this paper I have used the descriptive anatomical name ("premotor cortex," "rostro-ventral," etc.) to refer to the premotor cortex and its subsegments. For the sake of brevity, however, I have sometimes used the other nomenclature (area F5, etc.) as well.

[3] Two papers in 1996 first used the term *mirror neuron*: G. Rizzolatti, L. Fadiga, V. Gallese, and L. Fogassi, "Premotor Cortex and the Recognition of Motor Actions," *Cognitive Brain Research* 3 (1996): 131–41; V. Gallese, L. Fadiga, L. Fogassi, and G. Rizzolatti, "Action Recognition in the Premotor Cortex," *Brain* 119 (1996): 593–609.

[4] Christian Jarrett "Mirror Neurons: The Most Hyped Concept in Neuroscience? Mirror Neurons Are Fascinating but They Aren't the Answer to What Makes Us Human," Brain Myths, *Psychology Today*, December 10, 2012, http://www.psychologytoday.com/blog/brain-myths/201212/mirror-neurons-the-most-hyped-concept-in-neuroscience (accessed June 2, 2015).

[5] V. S. Ramachandran, "The Neurons That Shaped Civilization," TED talk, November 2009.

[6] The key papers are Ruth Leys, "'Both of Us Disgusted in My Insula': Mirror-Neuron Theory and Emotional Empathy," in *Science and Emotions after 1945*, ed. Frank Biess and Daniel M. Gross (Chicago: University of Chicago Press, 2014), 67–95 (an earlier version of the paper appeared at nonsite.org); Allan Young, "The Social Brain and the Myth of Empathy," *Science in Context* 25, no. 3 (2012): 401–24; Allan Young, "Mirror Neurons and the Rationality Problem," in *Rational Animals, Irrational Humans*, ed. S. Watanabe, A. P. Blaisdell, L. Huber, and A. Young (Tokyo: Keio University Press, 2009), 67–80; Susan Lanzoni, "Imaging Emotions: Reconfiguring the Social in Neuroscience," paper presented at the annual meeting of the American Association for the History of Medicine, May 8–11, 2014. Of course mirror neuron research has been critiqued from within the biomedical fields as well. I don't have the space to discuss this criticism here nor the interesting popular response to mirror neurons, but I plan to do so in my current book project, a history of mirrors in the mind sciences.

7   They did so as part of larger projects to critically examine the science of affect (Leys) and the social brain (Young).

8   Leys, "'Both of Us Disgusted,'" 72. Young agrees with Leys that mirroring is depicted as a direct process, a "brain-to-brain product, unmediated by mental states" ("The Social Brain and the Myth of Empathy," 403). Following the example of a number of philosophers, Leys and Young focus their analysis on "intentionality," which is a major element in the discussion of mirror neurons. See Emma Borg, "If Mirror Neurons Are the Answer, What Is the Question?," *Journal of Consciousness Studies* 14 (2007): 5–19; Pierre Jacob, "What Do Mirror Neurons Contribute to Human Social Cognition?," *Mind and Language* 23 (2008): 190–223; Shaun Gallagher, "Simulation Trouble," *Social Neuroscience* 2–3 (2007): 353–65. I do not address the question of intentionality directly here, but I submit that the argument could be extended to take it into account.

9   I pair autism and psychopathy here only because that juxtaposition was important for the scientists I study.

10  The mirror neuron researchers also refer to the experiments by Clinton Woolsey. After Penfield similar maps were constructed for other animals, including Woolsey's "simiusculus," a map of the motor and premotor areas of the rhesus macaque (*Macaca mulatta*). C. N. Woolsey, P. H. Settlage, D. R. Meyer, W. Sencer, T. Pinto Hamuy, and A. M. Travis, "Patterns of Localization in Precentral and 'Supplementary' Motor Areas and Their Relation to the Concept of a Premotor Area," *Research Publications—Association for Research in Nervous and Mental Disease* 30 (1952): 238–64. Woolsey's simiusculus was based on Penfield's representation. The medical historian Erwin Ackerknecht, Woolsey's colleague at the University of Wisconsin, suggested this simian equivalent to him. Four years earlier Woolsey had mapped the tactile areas of the rat's cerebral cortex: C. N. Woolsey and D. H. Le Messurier, "The Pattern of Cutaneous Representation in the Rat's Cerebral Cortex," *Federation Proceedings* 7 (1948): 137.

11  Note, however, that Penfield also produced combined sensory and motor homunculi to illustrate the sequential pattern of the map, for example in Wilder Penfield and Edwin Boldrey, "Somatic Motor and Sensory Representation in the Cerebral Cortex of Man as Studied by Electrical Stimulation," *Brain* 60, no. 4 (1937): 432. But it was clear that the somatosensory and motor areas were separate.

12  For a history of this approach within the larger tradition of localization of function, see my forthcoming book: Katja Guenther, *Localization and Its Discontents: A Genealogy of Psychoanalysis and the Neuro Disciplines* (Chicago: University of Chicago Press, 2015).

13  Giacomo Rizzolatti, *Mirrors in the Brain: How Our Minds Share Actions and Emotions* (Oxford: Oxford University Press, 2008), 3.

14 G. Rizzolatti, R. Camarda, L. Fogassi, M. Gentilucci, G. Luppino, and M. Matelli, "Functional Organization of Inferior Area 6 in the Macaque Monkey. II: Area F5 and the Control of Distal Movements," *Experimental Brain Research* 71 (1988): 503.

15 M. Gentilucci, L. Fogassi, G. Luppino, M. Matelli, R. Camarda, and G. Rizzolatti, "Functional Organization of Inferior Area 6 in the Macaque Monkey. I: Somatotopy and the Control of Proximal Movements," *Experimental Brain Research* 71 (1988): 476. The paper is the first part of a pair: Rizzolatti et al., "Functional Organization of Inferior Area 6 in the Macaque Monkey. II."

16 Gentilucci et al., "Functional Organization of Inferior Area 6 in the Macaque Monkey. I," 476.

17 Penfield thought that, in addition to the primary motor cortex, the supplementary motor cortex was organized somatotopically, but not the premotor cortex. For his somatotopic maps of the supplementary motor cortex, see Wilder Penfield and Herbert Jasper, *Epilepsy and the Functional Anatomy of the Human Brain* (Boston: Little, Brown, 1954), 105.

18 Gentilucci et al., "Functional Organization of Inferior Area 6 in the Macaque Monkey. I," 478. This was also indicated by listing more than one body part, for example "H, A" (hand, arm), or "H, M" (hand, mouth).

19 The researchers use the term *active movements* in the 1988 pair of papers (e.g., Gentilucci et al., "Functional Organization of Inferior Area 6 in the Macaque Monkey. I," 478). They later generally prefer *complex actions* or *motor acts* (e.g., Rizzolatti, *Mirrors in the Brain*, especially chapter 2).

20 Rizzolatti et al., "Functional Organization of Inferior Area 6 in the Macaque Monkey. II," 506.

21 It was this attribution of sensory properties to mirror neurons that was at the heart of much of the internal criticism of mirror neuron research. For an important and recent critique of this and related aspects, see Gregory Hickok, *The Myth of Mirror Neurons: The Real Neuroscience of Communication and Cognition* (New York: Norton, 2014), 64, chapters 4–8.

22 Gentilucci et al., "Functional Organization of Inferior Area 6 in the Macaque Monkey. I," 481.

23 Rizzolatti et al., "Functional Organization of Inferior Area 6 in the Macaque Monkey. II," 506.

24 Gentilucci et al., "Functional Organization of Inferior Area 6 in the Macaque Monkey. I," 487.

25 The area is located above F4 and F5. G. Rizzolatti, M. Gentilucci, R. M. Camarda, V. Gallese, G. Luppino, M. Matelli, and L. Fogassi, "Neurons Related to Reaching-Grasping Arm Movements in the Rostral Part of Area 6 (Area 6aβ)," *Experimental Brain Research* 82 (1990): 337–50.

26  This was an "inhibitory modulation," the absence of neuronal activity. Because this "inhibition" is distinct from the one I am interested in here, for the sake of clarity I have not discussed it. A further terminological overlap is *congruent* and *noncongruent*, which have been used to compare actions (the meaning I am interested in) but also to compare neuronal activity.

27  Note that the premovement modulation could be either of the same or of a different sign as the activity during movement; that is, it could be excitatory or inhibitory.

28  Rizzolatti also included trials during which the monkey was prevented from seeing its arm and the food by a plastic plane. Because of previous training, the monkey knew that the food was reachable, even if it couldn't be seen. The experiment aimed to rule out the possibility that neuronal activity during movement could be attributed to the visual stimulus. Rizzolatti et al., "Neurons Related to Reaching-Grasping Arm Movements in the Rostral Part of Area 6," 339.

29  Ibid., 347.

30  Among other arguments, the researchers pointed out that the activity of reaching-grasping neurons went beyond preparation of the specific movement; for example, they were "influenced by the distance of the stimulus from the monkey," a factor that introduced sensory processing into the mix (ibid., 347).

31  DiPellegrino et al., "Understanding Motor Events," 176.

32  Ironically the shift in the experiment and the emergence of mirroring as an explicit element of the discussion made the one-way mirror superfluous. If the stimulus was the movement of the scientist rather than a piece of food, it no longer made sense to stash it behind a piece of glass.

33  DiPellegrino et al., "Understanding Motor Events," 177.

34  Rizzolatti et al., "Premotor Cortex and the Recognition of Motor Actions," 135. See also Gallese et al., "Action Recognition in the Premotor Cortex," 600–602.

35  Gallese et al., "Action Recognition in the Premotor Cortex," 601.

36  Rizzolatti et al., "Premotor Cortex and the Recognition of Motor Actions," 135.

37  DiPellegrino et al., "Understanding Motor Events," 176.

38  This grew out of a collaboration between Rizzolatti's group at the Istituto di Fisiologia and the Center for Clinical Neurology in Parma. L. Fadiga, L. Fogassi, G. Pavesi, and G. Rizzolatti, "Motor Facilitation during Action Observation: A Magnetic Stimulation Study," *Journal of Neurophysiology* 73, no. 6 (1995): 2608–11. In the study the researchers measured so-called motor evoked potentials (MEPs) from the hand muscles of their human

subjects. The MEPs were triggered through simultaneous application of transcranial magnetic stimulation (TMS) to the motor areas of the brain. Then, in order to explore the "correlation" between observed and executed action, in an additional trial in the same study they recorded the electromyogram activity of the set of muscles they had explored in the first experiment, while the subjects were executing grasping and other movements. In other words, the researchers compared brain activity measured through MEP and muscular activity measured through electromyography to compare observation and execution. They found that the MEP pattern "reflected the pattern of muscle activity" in the electromyogram experiment and that they were "very similar" (2608, 2609).

39  V. S. Ramachandran, "The Neurology of Self-Awareness," *Edge*, January 8, 2007, http://edge.org/3rd_culture/ramachandran07/ramachandran07_index.html (accessed June 2, 2015). See also V. S. Ramachandran, *The Tell-Tale Brain: A Neuroscientist's Quest for What Makes Us Human* (New York: Norton, 2011).

40  V. S. Ramachandran, "Mirror Neurons and Imitation Learning as the Driving Force behind the Great Leap Forward in Human Evolution," *Edge*, May 31, 2000, https://edge.org/conversation/mirror-neurons-and-imitation-learning-as-the-driving-force-behind-the-great-leap-forward-in-human-evolution/ (accessed June 2, 2015).

41  For the full theory, see ibid.

42  Ibid.

43  In one experiment one group of participants was asked to think of college professors while responding to a set of "general knowledge" questions, whereas the other group was asked to think of soccer hooligans. The first group clearly outperformed the second in the task, from which Dijksterhuis concluded, "Imitation can make us slow, fast, smart, stupid, good at math, bad at math, helpful, rude, polite, long-winded, hostile, aggressive, cooperative, competitive, conforming, nonconforming, conservative, forgetful, careful, careless, neat, and sloppy." Quoted in Marco Iacoboni, *Mirroring People: The Science of Empathy and How We Connect with Others* (New York: Farrar, Straus and Giroux, 2009), 200–201.

44  Ibid., 202. Rizzolatti argued along similar lines for "increasingly articulated and complex mirror neuron systems" in humans (*Mirrors in the Brain*, 192).

45  For an account of the collaboration, see Iacoboni, *Mirroring People*, 192–95.

46  Roy Mukamel, Arne Ekstrom, Jonas Kaplan, Marco Iacoboni, and Itzhak Fried, "Single-Neuron Responses in Humans during Execution and Observation of Actions," *Current Biology* 20 (2010): 750–56.

47  Iacoboni, *Mirroring People*, 193.

48  Mukamel et al., "Single-Neuron Responses in Humans during Execution and Observation of Actions," 753, my emphasis.

49  For this reason they have also been called "anti-mirror neurons." C. Keysers and V. Gazzola, "Social Neuroscience: Mirror Neurons Recorded in Humans," *Current Biology* 20 (2010): R353–R354.

50  Mukamel et al., "Single-Neuron Responses in Humans during Execution and Observation of Actions," 755.

51  Iacoboni, *Mirroring People*, 220.

52  G. Rizzolatti and M. A. Arbib, "Language within Our Grasp," *Trends in Neurosciences* 21, no. 5 (1998): 188–94.

53  There is a long history of debates over the origins of language. See Gregory Radick, *The Simian Tongue: The Long Debate about Animal Language* (Chicago: University of Chicago Press, 2007).

54  Rizzolatti and Arbib, "Language within Our Grasp," 190.

55  With this Rizzolatti and Arbib drew on the motor theory of speech perception that was formulated in the 1950s by the psychologist Alvin Liberman. For a history and critique of the motor theory of speech perception, see Hickok, *The Myth of Mirror Neurons*, especially chapter 5.

56  Rizzolatti and Arbib, "Language within Our Grasp," 190.

57  Ibid., 191, 193.

58  M. Gentilucci, F. Benuzzi, M. Gangitano, and S. Grimaldi, "Grasp with Hand and Mouth: A Kinematic Study on Healthy Subjects," *Journal of Neurophysiology* 86 (2001): 1685–99. In 2003 they found similar responses when subjects were merely observing grasping movements. M. Gentilucci, "Grasp Observation Influences Speech Production," *European Journal of Neuroscience* 17 (2003): 179–84.

59  Rizzolatti and Arbib, "Language within Our Grasp," 193.

60  *Empathy* is the English translation of the German *Einfühlung* (a term introduced in 1903 by Theodor Lipps), which entered American academic psychology in the first decade of the twentieth century in the work of Edward Titchener. Susan Lanzoni, "Empathy in Translation: Movement and Image in the Physiology Laboratory," *Science in Context* 25 (2012): 301–27.

61  B. Wicker, C. Keysers, J. Plailly, J.-P. Royet, V. Gallese, and G. Rizzolatti, "Both of Us Disgusted in *My* Insula: The Common Neural Basis of Seeing and Feeling Disgust," *Neuron* 40 (2003): 655, 660.

62  Ibid., 655, 660–61.

63  Of course asking them would have broken with the theory's assumption of direct processing of emotions, as verbal communication about someone's emotional state is possible only via a cognitive route, as Leys points

out. The researchers tried to fix the problem in a follow-up study, but, as Leys argues, not successfully. Leys, "'Both of Us Disgusted.'"

[64] Mbemba Jabbi, Marte Swart, and Christian Keysers, "Empathy for Positive and Negative Emotions in the Gustatory Cortex," *NeuroImage* 34, no. 4 (2007): 1744–53.

[65] Leys, "'Both of Us Disgusted,'" 79.

[66] Young also analyzes pathologies of empathy (autism and psychopathy). But because he remains within the paradigm of empathy and its seeming effacement of difference, he focuses mostly on the aversion of the scientists to accept the possibility of nonsocial versions of empathy. Allan Young, "Empathic Cruelty and the Origins of the Social Brain," in *Critical Neuroscience: A Handbook of the Social and Cultural Contexts of Neuroscience*, ed. Suparna Choudhury and Jan Slaby (Blackwell, 2012), Blackwell Reference Online.

[67] H. Meffert, V. Gazzola, J. A. den Boer, A. Bartels, and C. Keysers, "Reduced Spontaneous but Relatively Normal Deliberate Vicarious Representations in Psychopathy," *Brain: A Journal of Neurology* 136 (2013): 2550–62.

[68] Christian Keysers, *The Empathic Brain: How the Discovery of Mirror Neurons Changes Our Understanding of Human Nature* (N.p.: Social Brain Press, 2011), 206.

[69] As Keysers points out in his blog, the scientists moved their subjects to their scanning facility "one by one, in bullet proof minivans." Because no metal could be brought near the magnetic imaging scanner, the guards did not carry any firearms, but the patients "had wooden sticks sewn into their trousers and plastic hand-cuffs to keep them from running away or hurting anyone." Christian Keysers, "Inside the Mind of a Psychopath: Empathic, but Not Always," Empathic Brain, *Psychology Today*, July 24, 2013, psychologytoday.com/blog/the-empathic-brain/201307/inside-the-mind-psychopath-empathic-not-always (accessed June 2, 2015).

[70] The avoidance of the term *mirror neurons* is consistent with the decline of interest in mirror neuron research after the "period of most intense interest between 2002 and 2007" (Young, "The Social Brain and the Myth of Empathy," 403).

[71] Meffert et al., "Reduced Spontaneous but Relatively Normal Deliberate Vicarious Representations in Psychopathy," 2552, 2559.

[72] Keysers, "Inside the Mind of a Psychopath." To the researchers this suggested possibilities for treatment (Meffert et al., "Reduced Spontaneous but Relatively Normal Deliberate Vicarious Representations in Psychopathy," 2560.

[73] It also offered new possibilities for treatment, as Meffert et al. pointed out: therapies could "harvest the patients' potential to normalize vicarious activations through deliberately focusing attention on empathizing

with others" ("Reduced Spontaneous but Relatively Normal Deliberate Vicarious Representations in Psychopathy," 2562).

74 TMS of the motor cortex evokes an electrical potential in the corresponding muscle, called a motor evoked potential (MEP).

75 S. Fecteau, A. Pascual-Leone, and H. Théoret, "Psychopathy and the Mirror Neuron System: Preliminary Findings from a Non-psychiatric Sample," *Psychiatry Research* 160 (2008): 143.

76 Ibid., 142. See also the earlier distinctions in the paper between forms of empathy, for example by R. J. Blair between cognitive, motor, and emotional empathy (143).

77 Ibid., 142.

78 The subjects were asked to imitate lip forms presented to them in still pictures. Nobuyuki Nishitani, Sari Avikainen, and Riitta Hari, "Abnormal Imitation-Related Cortical Activation Sequences in Asperger's Syndrome," *Annals of Neurology* 55 (2004): 558–62.

79 L. Oberman, E. Hubbard, J. McCleery, E. Altschuler, V. S. Ramachandran, and J. Pineda, "EEG Evidence for Mirror Neuron Dysfunction in Autism Spectrum Disorders," *Cognitive Brain Research* 24 (2005): 190–98. Ramachandran first presented results from EEG experiments at the annual meeting of the Society for Neuroscience in 2000. E. L. Altschuler, A. Vankov, E. M. Hubbard, et al., "Mu Wave Blocking by Observation of Movement and Its Possible Use to Study the Theory of Other Minds," *Society for Neuroscience* (2000), abstracts 68.1.

80 Oberman et al., "EEG Evidence for Mirror Neuron Dysfunction in Autism Spectrum Disorders," 195; V. S. Ramachandran and Lindsay M. Oberman, "Broken Mirrors," *Scientific American* 295 (2006): 63–69.

81 J. Williams, G. Waiter, A. Gilchrist, D. Perrett, A. Murray, and A. Whiten, "Neural Mechanisms of Imitation and 'Mirror Neuron' Functioning in Autistic Spectrum Disorder," *Neuropsychologia* 44 (2006): 610–21.

82 For example by Simon Baron-Cohen and Uta Frith on theory of mind, by Alyson Bacon on empathy, and by Kjelgaard on language: S. Baron-Cohen, "Theory of Mind and Autism: A Review," in *International Review of Research in Mental Retardation*, vol. 23: *Autism*, ed. L. M. Glidden (San Diego: Academic Press, 2001), 169–84; U. Frith, *Autism: Explaining the Enigma* (Oxford: Blackwell, 1989); A. Bacon, F. Fein, R. Morris, L. Waterhouse, and D. Allen, "The Responses of Autistic Children to the Distress of Others," *Journal of Autism and Developmental Disorders* 2 (1998): 129–42; M. Kjelgaard and H. Tager-Flusburg, "An Investigation of Language Impairment in Autism: Implications for Genetic Subgroups," *Language and Cognitive Processes* 16 (2001): 287–308.

83 J. Williams, A. Whiten, T. Suddendorf, and D. Perrett, "Imitation, Mirror Neurons and Autism," *Neuroscience and Biobehavioral Reviews* 25 (2001): 287–95.

84 Ibid., 289, 291.

[85] P. D. McGeoch, D. Brang, and V. S. Ramachandran, "Apraxia, Metaphor and Mirror Neurons," *Medical Hypotheses* 69 (2007): 1167. David Brang, who is now at Northwestern University, was at the time a doctoral student of Ramachandran's.

[86] Oberman et al., "EEG Evidence for Mirror Neuron Dysfunction in Autism Spectrum Disorders," 196.

[87] Fecteau et al., "Psychopathy and the Mirror Neuron System," 142.

[88] Note that the relationship wasn't always as clear. In a letter to the editor in response to Meffert et al.'s paper, Oberman and colleagues pointed out that in autism too there was a "dissociation between spontaneous and deliberate experience." For example, individuals with Asperger's syndrome, a condition closely related to autism, were able to "understand mental states such as desires and beliefs (mentalizing) when explicitly prompted to do so" even though they showed difficulty in doing so spontaneously. S. Gillespie, J. McCleery, and L. Oberman, "Spontaneous versus Deliberate Vicarious Representations: Different Routes to Empathy in Psychopathy and Autism," *Brain: A Journal of Neurology* 137 (2014): 1.

[89] Wicker et al., "Both of Us Disgusted in *My* Insula," 661.

[90] Wicker's assumption of "basic emotions" can be traced back to the work of American psychologists Silvan S. Tomkins and Paul Ekman (Leys, "'Both of Us Disgusted,'" 73).

[91] See also Alvin Goldman, *Simulating Minds: The Philosophy, Psychology, and Neuroscience of Mindreading* (Oxford: Oxford University Press, 2006). TT is often lumped together with the broader "theory of mind" of explaining the understanding of the outside world, and in particular the mental states of others. I use "theory of mind" in a broader sense here: it can be ST or TT. For a short history of TT, see Iacoboni, *Mirroring People*, 168–69.

[92] Vittorio Gallese and Alvin Goldman, "Mirror Neurons and Simulation Theory of Mind-Reading," *Trends in Cognitive Science* 2, no. 12 (1998): 493. Theory theory was first formulated in the 1980s in developmental psychology. See Margaret Boden, *Mind as Machine: A History of Cognitive Science* (Oxford: Oxford University Press, 2006).

[93] Williams et al., "Imitation, Mirror Neurons and Autism," 288.

[94] V. Gallese, "The 'Shared Manifold' Hypothesis: From Mirror Neurons to Empathy," *Journal of Consciousness Studies* 8 (2001): 45.

*Tobias Rees*

# 10 On How Adult Cerebral Plasticity Research Has Decoupled Pathology from Death

PROCHIANTZ WAS WATCHING outside a window, overseeing a Paris covered in pigeon dirt. I looked at my scribbles. "Death," I had noted him saying, "is a solution. Life is a problem."

Let me explain.

## A Silent Embryogenesis

PARIS, OCTOBER 2002, in the office of Alain Prochiantz, professor at the École Normale Supérieure.

"What if cell death were normal?"

Prochiantz looked at me, provocatively. We were sitting in his office, a small box-shaped space on the seventh floor of the natural science building of the École Normale Supérieure, where he ran a neurobiology lab.

I wondered what to do with his question. Cell death normal? The death of neurons, which has such devastating consequences for humans? Normal?

A few days earlier I had by chance overheard a coffee conversation between two of Prochiantz's senior researchers, Alain Trembleau and Michel Volovitch, who discussed whether they would invest in a biotech company he was about to launch.

I was curious to find out what the company was all about and, after a day of hesitation, I approached Prochiantz and asked him if he would mind telling me about his startup. Had it anything to do with his research on brain plasticity?

Since February I had been studying his lab's effort to think about the adult human brain in embryogenetic terms—an effort that to many of his colleagues was an outrageous provocation.[1] The history of this provocation dates back to 1989.

At that time Prochiantz was working as an embryologist of the central nervous system and was interested in the relevance of the recently discovered homeotic genes for cellular morphogenesis.[2] It had been discovered in the mid-1980s that homeotic genes are preserved across phyla and organize the emergence of a basis body axis. Prochiantz was curious whether perhaps homeoproteins—transcription factors coded for by homeotic genes—were critical not only for the embryogenetic formation of a basic body axis but also for the migration and differentiation of cells. Perhaps homeoproteins direct a given cell to become a cell of the brain instead of the toe? And once in the brain a cell of the dentate gyrus rather than of the olfactory bulb?

In a control experiment his graduate student Alain Joliot had added homeoproteins—coded for by homeotic genes—to the extracellular milieu of mature neurons in a Petri dish. The aim of the control was to show that the mere co-culturing of homeoproteins and neurons had no effect on neurons. However, the next day Joliot and Prochiantz found the homeoproteins in the nuclei of the cells where they seemed to have caused an intense morphogenetic outgrowth.

Could it be that homeoprotein can travel between neurons and cause—or allow for—morphogenetic changes?

Implausible that possibility seemed for several reasons.

First, cell biologists had firmly established that cells are autonomous. The assumption was that each cell has a genome, and this genome—if modified by signaling molecules received from other cells—determines what will happen in a cell. The control experiment of Prochiantz and Joliot, however, seemed to suggest that a whole protein, a transcription factor, could enter a cell, go straight to the nucleus, and activate the genes responsible for morphogenesis. Could that be? When they explored this possibility with colleagues in other labs, the straightforward answer they got was *no*.

In Joliot's words, "Not only would such a transfer have violated the principle of cell autonomy––it would also have undermined everything that was known about molecular signaling: How could something as big as a transcription factor leave a nucleus, travel through the cytoplasm, leave the cell, find its way through the extracellular milieu, and enter another cell, and all of this without degeneration of the protein?"

Second, at least since the 1880s histologists had known that humans are born with a definite number of nerve cells, and at least since the 1890s that the spectacular growth of the fine structure of these nerve cells comes to end once maturity is reach.[3]

"In fully grown animals," as Ramón y Cajal wrote around the late 1890s, "the nervous system is essentially fixed."

A morphogenetic outgrowth in already mature neurons thus ran counter to a century of careful neuroanatomical and neurophysiological research and hence seemed somewhat unlikely.

Despite—or because of—the provocative implications of their observation, however, Prochiantz felt they may have made a major discovery. Further experiments followed and not only established that the nonautonomous transfer of homeoproteins was a regularly occurring event, at least in vitro. They also showed that homeoproteins were actually expressed in a whole series of animals, including humans, up until adulthood.

Perhaps one has to pause for a moment to understand the surprise this was: What actually would homeoproteins, that is, embryogenesis-specific transcription factors that control the cellular emergence of form where before there was none, do in adult human brains? Especially as adult human brains were supposed to be fixed and immutable cellular structures?

The answer—at the time entirely speculative—that Prochiantz and Joliot came up with was that homeoproteins are perhaps cell-independent plastic forces that, by way of transferring between neurons, activate genes that render a cell plastic, on the level of its form as much as on the level of its connections.

In 1991 they went public, publishing a short paper in the *Proceedings of the National Academy of Sciences* in which they reported that they had discovered a yet unknown embryogenetic signaling mechanism: the noncell autonomous transfer of homeoproteins; that this signaling mechanism seemed critical for neuronal morphogenesis; and that it also occurred in adult human brains, where it rendered mature cells morphogenetically plastic.[4]

The reactions they got from their paper were overwhelmingly negative. Their colleagues were adamant that proteins do not slide between cells and that, as had been known and shown for over a century, the brain is a fixed and immutable cellular tissue. Prochiantz, however, did not let go. As if unmoved by the critique of his peers, he continued to insist—in scientific publications as well as in a series of popular science books and talks— that he and Joliot had made an important discovery, one that might overthrow neurobiology as we knew it.[5] His colleagues, outraged by Prochiantz's perseverance, began to turn away from him. Within a year or two his lab had been pushed to the margins of the Parisian community of neuroscientists, where it continued to work on what seemed ridiculous to those who surrounded them: the retained embryogenetic plasticity of the adult human brain.[6] And then, after almost ten years of polemic and provocation, the unexpected happened. In the late 1990s two American laboratories began to publish papers in which they reported that every day thousands of neurons are born in the brains of adult humans and nonhuman primates.[7] The effect of these publications on neuroscience was that of an unanticipated—and thus turbulent—conceptual opening. The reports on adult neurogenesis quickly put in question the century-old truth of adult cerebral fixity; however, it undermined the established truth without providing a new conceptual framework for how to think about the brain. After all, no one knew yet whether the brain would be generally plastic, or, for that matter, what generally plastic would actually mean.[8]

For Prochiantz the discovery of adult neurogenesis amounted to a sudden change in the perception of his lab's research. To many it suddenly seemed as if he and his colleagues had elaborated an answer to a question that was only now being asked: Is the brain plastic? Are basic embryogenetic processes continuing in the adult human brain? Which molecules regulate adult cerebral plasticity? After the turn of the millennium Prochiantz thus could emerge as one of the internationally most renowned plasticity researchers and spokesperson of a plastic conception of the adult human brain.

What had until recently been utter nonsense had become avant-garde.[9]

"What," Prochiantz repeated his question, "if cell death were normal? What if it were a normal process, a way of getting rid of sick, old, superfluous cells? The idea of the company is just this, that perhaps the problem is less cell death than cell life."

"Can you explain more?" I asked.

He sighed but continued. "For a long time the diseases of the brain were thought of in terms of death. Axons and dendrites dry up; neurons die. That is also why we speak of neurodegenerative diseases. What, though, if the death of neurons or axons and dendrites were a normal physiological process? What if the problem of many diseases actually were not degeneration but the insufficient birth of new neurons or of new cellular tissue? Plasticity, I think, opens up a whole new and unexpected way for thinking about cerebral pathologies."

"And your plan is to find out if homeoproteins control the birth of new neurons in the adult?"

Prochiantz smiled. "Isn't that what they also do in the embryo?"

"Wouldn't that imply that embryology is the science proper to the adult brain?"


## After Death?

WHEN I LEFT Prochiantz's office I was tremendously excited.

It was as if I could suddenly see a possibility.

Wasn't his talk about the impact of plasticity on pathology suggesting that plasticity caused a conceptual opening of pathology? That a new conceptualization of pathology was emerging, one that undermined the conceptualization of pathology that was contingent on the older, the fixed and immutable brain? Wasn't he suggesting that plasticity was a crack in the reality of disease? Wasn't he implying that his research—plasticity research—would shatter what disease was and give rise to a radically new concept of disease and pathology?

What excited me as well was the empirical possibility.

Wouldn't it be possible for me, by way of following the work of his company, to study the experimental and conceptual labor that was necessary for the emergence of this new disease concept? What would disease be if it were no longer a matter of cell death? Could disease—could pathology—move "after death"? How? What new, what other concept of disease would emerge?

Over the following weeks and months, as I followed the work of Prochiantz's lab, the question if, and if then how, pathology could be decoupled from death seemed to me more and more an invitation to historically explore when disease was first thought of as death.

When actually were the first pathological studies of the brain published? Who actually cut his way through cerebral tissue and began correlating lesions and symptoms? And when did the first cellular, the first neuronal pathologies of the brain emerge? Who enrolled the brain in cell theory and who cellularized pathology? When and by whom was neuronal pathology explicitly linked to cell death?

And most significant with respect to my fieldwork, was the observation that the adult brain is not—on the level of cells—plastic an important episode in the history of cerebral pathology?

## The Emergence of (Cellular) Cerebral Pathology (ca. 1820s to 1870s)

THE FIRST PATHOLOGIES of the brain were published in the mid-1820s by the French physician Jean Baptiste Bouillaud (1796–1881), who was among the first to be educated into the new conception of disease introduced by Xavier Bichat (1771–1802).[10]

Up until the late 1790s physicians had assumed that disease was the result of a troubling of the four humors. It may sound curious, yet the modern conception of medicine, which assumes that what is underlying a disease are distinct pathophysiological processes that cause similar symptoms in different patients, was unknown for most of European history.[11] And it was largely the work of Bichat that changed this.[12]

Beginning in the late 1790s Bichat argued that his anatomical studies had convinced him that the challenge of medical research was not to think about humors but to correlate the symptoms of individual cases with postmortem tissue analysis in such a way that one could gradually work out a systematic anatomical understanding of disease.[13]

Bichat died young, and he published only three books. Yet already by the 1810s (when Bouillaud entered medical school) his studies had changed what medicine was about.[14] To be sick now no longer meant to have an imbalance in one's humors but rather to suffer from the degeneration of tissue, which Bichat and his successors—in the absence of any knowledge about infectious diseases—understood as the victory of death over life.[15]

Pathology, as it emerged in the work of Bichat between 1799 and 1801, was the science that maps and classifies the traces of death in the body and understands these traces as the actual cause of disease.

The significance of the work of Bouillaud was that he applied Bichat's pathology to *Homo cerebralis*, as it had emerged in the work of Franz Josef Gall.[16]

Gall (1758–1828), a Viennese physician who was working in prisons and asylums, had argued since the 1790s that the brain was the organ of the human. To be more precise, he claimed that his research led him to identify the exactly twenty-seven task-specific regions of the brain and that he could map the brain's functional geography onto the cranium and thus draw conclusions, by way of analyzing the shape of the skull of a given person, about that person's character and intellectual capacities.[17]

In 1807 Gall left Vienna for Paris, where he gave several talks about his craniology at the Académie Française. In the audience at one of these lectures was Bouillaud, who subsequently began to wonder whether he couldn't provide the pathological proof—against Jean Pierre Flourens (1794–1867), Gall's major Parisian opponent—that the brain is indeed composed of task-specific regions.

Wouldn't it be possible to find, for example, speech-impaired patients and compare their brains after their death?

Bouillaud began to look out for patients with speech loss and then systematically correlated his clinical observations with post-mortem anatomical dissection. What he discovered, around 1824, was that a whole series of patients with speech loss had a lesion of the left frontal lobe. Speech, he concluded, was located in the frontal lobe.

Bouillaud's pathological studies thus correlated two previously separate lines of argument: on the one hand Bichat's pathology;

on the other hand the argument that the human is a product of the brain. With Bouillaud the idea emerges that where brain tissue is impaired, the human is impaired as well.[18]

While Bouillaud sought in vain to convince his Parisian peers that the brain is organized in function-specific areas—an argument that was accepted only in the early 1860s, thanks to the studies of Paul Broca (1824–80)—a new understanding of organisms emerged in Germany that was to lastingly modify Bichat's conception of pathology. Beginning in the 1830s Matthias Schleiden (1804–81) and Theodor Schwann (1810–82) began to argue that and to explore whether all known organisms, indeed all of life, is made up of cells. Already by the 1850s their successors and followers had established that even the nervous system is no exception to this general rule: the brain, as any other organ (as any other plant, as Schleiden had it), was composed of cells.[19]

The conclusion—for many mid-nineteenth-century physicians an incredible one—was that one could now study cells and make conclusions about the human.

Among the first to apply cell theory to brain pathology were Robert Remak (1815–65) and Rudolf Virchow (1821–1902), both working in Johannes Müller's lab in the 1840s and early 1850s (where Schleiden and Schwann had worked as well). Over the next decade their work in the lab and the dissection room led both Remak and Virchow, roughly at the same time, to come up with the claim that all diseases, whether of the brain or of other organs, were due to and had to be found, indeed could only be found on the cellular level: what causes disease is necrosis, the degeneration and eventual death of cells.[20]

However, while Remak's Jewish background prevented him from becoming a professor, Virchow made a career.[21] In 1859 he went ahead and published, without much reference to Remak, his *Cellularpathologie*, the first sustained and systematic argument that medical, pathological research ought to be grounded in the cellular study of tissue.[22]

While Virchow understood his work as an event in the history of medicine, he was adamant that it was neither a critique of nor a departure from Bichat. It was just that, or so he writes, the work of Schwann, combined with his clinical and laboratory work, led him to think that it was no longer enough to give a mere description of tissue degeneration. Progress had occurred and had shown that tissue was cellular. Accordingly it was now necessary to use the microscope and to map disease on the level of cells. That this would partly change what medicine was about was inevitable and not a critique of Bichat. For Virchow, thus, disease was still the trace of death. It is just that the level on which one now had to map the trace of death had become cellular.

At least in principle the diseases of the brain became with Virchow and Remak a matter of cell degeneration (necrosis). I write "in principle" because Virchow actually had little to say about cerebral diseases. His lectures on cell pathology were much less concerned with an actual cell pathological study, whether of the brain or of other tissue, than with proving a principle: that all tissue is made up of cells. It took another one and a half decades for cerebral diseases to become neurodegenerative diseases. The reference here is largely to Theodor Meynert and to Jean-Martin Charcot.[23]

### Pathology and (the Absence of) Plasticity (1897 to 1914)

AROUND THE LATE 1870s brain research underwent a major mutation. Roughly at the time Meynert and Charcot published the first cell pathological studies of the brain, the discovery of staining techniques—most famously by Camillo Golgi—that allowed coloring of a single nerve cell of any one section of the brain opened up a whole new field of research: histology or the effort to understand the cellular makeup of the brain.[24]

Perhaps the most significant among the many significant events that shaped the nascent discipline of histology—one thinks, aside from Golgi, of Albrecht von Koelliker, Wilhelm Waldeyer, Wilhelm His, Richard Altmann, or Auguste Forel—were the studies of Santiago Ramón y Cajal (1852–1934), a Spaniard working largely in Madrid.[25] In the early 1890s, after he had spent several years studying the embryogenetic formation of the brain in chicken, cats, and dogs, Ramón y Cajal embarked on a large-scale, almost industrial project that kept him busy for a decade: a study of the cellular emergence of the brain in its entirety. Equipped with little more than a scalpel and his silver nitrate staining adapted from Golgi, he cut his way through the brains of various animals and humans of different ages and gradually reconstructed from his stains the cellular emergence of the brain, from conception to death. At the end of the 1890s he published the first of what would eventually be a two-volume histological atlas of the brain, *Textura del sistema nervioso del hombre y de los vertebrados*, which is still one of the major sources of today's comprehension of the central nervous system.[26]

The *Textura* was organized around a series of spectacular discoveries. The most significant of them were that nerve cells emerge only over the course of embryogenesis; that the fine structure— he spoke of arborization—grows until adulthood is reached; that nerve cells were and remained contiguous with one another (that is, the fine structure of one gets close to but never touches the fine structure of another one); and that once the threshold to maturity was passed, all cellular growth came to an end. The brain of adult animals is, as he put it in 1897, "essentially fixed."

In his interpretation of these findings, Ramón y Cajal drew two major conclusions that would give form to twentieth-century brain research.

The first was that the fine structure—previously hardly considered central for understanding the nervous system—had to be a key for understanding the brain. Not that no one had studied the fine structure before him. But the question that concerned most histologists at the time with respect to the fine structure was whether or not cells are contiguous with one another. It had been the dominant assumption since the 1870s (and the work of Joseph Gerlach) that the fine structure, which is fully in place before birth, connects cells with one another and merely grow bigger. However, since the 1880s a few researchers—notably His and Forel—had begun to argue that axons and dendrites grow freely from cells and do not seem to touch the axons and dendrite of other cells. Ramón y Cajal, however, argued what no one had argued before: that the free growth (arborization) of the fine structure of nerve cells had to be understood as the direct correlate of learning and memory (hence the vehemence with which he argued for contiguity).[27]

The second conclusion Ramón y Cajal drew was that only the young, the growing brain was plastic. Given that the growth of the fine structure had to be understood as the cellular substrate of learning, and given that this growth comes to an end once maturity is reached, it seemed only consequential to assume that the adult human brain was devoid of plastic potential. He was adamant that adult brains too could learn. However, there simply was not much space left for growth, except perhaps, or so he speculated at the end of his life, on the level of dendritic spines (which he discovered in the 1880s). Though any such potential for learning, he added in an almost somber tone, would be minimal compared to the plastic potential of the young, still growing brain.[28]

And pathology?

In 1905 Ramón y Cajal began to study the regeneration of neurons. Like many of his contemporaries, he cut through live nerve fibers in order to see whether the interrupted connection would regrow, whether a severed axon would emerge, or whether an injured nerve cell would die and a new one would emerge. In 1913–14, then already a towering figure of the cellular study of the brain (he had received the Nobel Prize in 1906), Ramón y Cajal published *Regeneration and Degeneration of the Nervous System*, a book-length study of his findings.

"A vast series of anatomic-pathological experiments in animals," he informs his readers early on, "and an enormous number of clinical cases that have been methodically followed by autopsy [have convinced me that it is] an unimpeachable dogma [that] there is no regeneration of the central paths, and [that] there is no restoration of the normal physiology of the interrupted conductors."

Sure, he concedes, "it has been demonstrated, beyond doubt, that there is a production of new fibers and clubs of growth in the spinal cord of tabetics . . . and of cones and ramified axons in the scar of spinal wounds of man and animals." However, he goes on, "these investigations, while they have brought out unquestionable signs of repair, which are comparable in principle with those of the central stump of the nerves, have also confirmed the old concept of the essential impossibility of regeneration, showing that . . . the restoration is paralyzed, giving place to a process of atrophy and definite break-down of the nerve sprouts." At the end of the book, after many hundred pages of erudite argumentation, Ramón y Cajal then also provides a conceptual explanation for why there actually could be no generation of cells or cellular outgrowth: "The functional specialization of the brain imposed on the neurones two great lacunae; proliferative inability and irreversibility of intraprotoplasmatic differentiation. It is for this reason that, once the development was ended, the founts of growth and regeneration of the axons and dendrites dried up irrevocably."[29]

Here Ramón y Cajal correlates his earlier studies of the cellular becoming of the brain to his later regeneration studies and explains the findings of the latter with the findings of the former: it is the functional specialization achieved during the plastic period that requires a more or less fixed and immutable and unchanging brain. Would there be new neurons, or merely new connections (intraplasmic differentiation), it would alter the specific pattern of arborization that is the result of an always singular life history—and this would upset not only the brain but the human.

Ramón y Cajal did not change the cell pathological theory of disease as it had emerged with Bichat and Virchow and Remak. However, he provided a powerful histological argument as to why pathology, when it comes to the brain, had to be and could only be concerned with cell degeneration and death. With Ramón y Cajal a powerful plasticity-pathology axis had thus emerged, according to which it is the absence of plasticity in the adult that defines the horizon of the pathological: precisely because there are no new cells, because there is no plasticity, pathology has to be concerned with cell death.

## A New Concept of Plasticity and Pathology (1960s)

HOW, OVER THE course of the short twentieth century, did the neuronal sciences mutate the plasticity-pathology axis formulated by Ramón y Cajal?

Curious as it may sound, they did not actually mutate it. Ramón y Cajal's dictum that the mature brain is devoid of growth processes and that neurons do not regenerate remained in place throughout the twentieth century. To point out a surprising conceptual continuity is not to deny that over the course of one hundred years spectacular technical innovations occurred (say, single-cell recording or the electron microscope) and that a large series of experimental observations have significantly refined (mutated) the neuronal comprehension of the brain (say, the discovery of the chemical nature of synaptic communication and the subsequent reconfiguration as chemical, metabolic, machines). However, it seems as if none of these innovations and observations (or mutations) has actually challenged the conceptual grid first worked out by Ramón y Cajal. From roughly 1900 to roughly 2000, no matter whether one looks at early twentieth-century electrophysiology, at mid-twentieth-century cybernetics, or at late twentieth-century neurochemistry, it was the absence of cellular plasticity that defined the horizon of the pathological.

The emergence of new concepts of plasticity and pathology in the mid-1960s did little to change this.

## A Chemical, Synaptic Brain Emerges

UP UNTIL WORLD War II, the neuronal study of the brain was extraordinarily heterogeneous. The brain of Ramón y Cajal, that is to say, was only one conceptualization among many others. Up north, for example, histology never really arrived. In France brain research meant largely clinical pathology and ablation experiments in the tradition of Bouillaud, Broca, and Charcot. Even farther north, in the United Kingdom, brain research took yet another form. To scientists like Charles Scott Sherrington, Keith Lucas, and Edgar Adrian the brain was an electrical organ, and the question to be solved was how sensory information is converted into electrical information and travels from the periphery to the brain and back. (Sherrington's famous answer in 1897 was that a "synapsis"—an electrical reaction—would occur in the gap between axon and dendrite and would thus integrate information from the peripheral nervous system and transmit it to the central nervous system.) And in Germany the emergence of the cellular brain since the 1840s had gradually given rise to cytoarchitecture, that is, the study of the brain in terms of cellular form and the kind of function different forms code for. (One thinks of the studies of Theodor Meynert, Oskar and Cecile Vogt, and Korbinian Brodmann.)[30]

In the aftermath of World War II this heterogeneity of brains suddenly and quickly dissolved. Within a mere decade, from roughly the mid-1940s to the mid-1950s, three separate and yet closely intertwined events gave rise to a somewhat unified and previously unknown understanding of the brain as a chemical, synaptic machine.

The first of these events was what one could call the *global-ization of the problem of the synapse*. Throughout the first half of the twentieth century Sherrington's concept of the synapse was only of local relevance. Outside the confines of his own lab few picked up on the synapse, whether in Europe or the United King-dom. (Not even Adrian picked up on the synapse, and even Sher-rington seems to have doubted that the whole brain is synaptic.) After the war, though, Sherrington's many students and postdocs secured important professorships in the United States, in Europe, and in Australia. What is more, several of them were successful in launching major new research centers, each one of which was concerned with the problem of the synapse. (To name but a few: Bernhard Katz in London, Alexander Forbes at Harvard, John Eccles in Canberra, Hodgkin and Huxley at Cambridge, Alfred Fessard in Paris, Wilder Penfield in Montreal, Giuseppe Moruzzi in Pisa, and Frédéric Bremer in Brussels.)[31] The synapse thus left Sherrington's lab and quickly came to define cutting-edge brain research in the world's most significant and cutting-edge brain research centers.

Second, in the late 1940s and early 1950s the synapse was *chem-icalized*. For Sherrington and his students the synapsis had been an electrical reaction; it occurred in the gap between dendrite and axon, and they studied how this reaction integrated infor-mation and made it meaningful. Gradually, though, the work of Katz, Feldberg, Kuffler, Fessard, and Eccles established that Sher-rington and his followers had gotten something wrong: synaptic communication, as they called it (thereby shifting the focus away from the synapsis and toward cell-to-cell communication), is not, as Sherrington had assumed, an electrical reaction—it is a chemi-cal event.[32]

And third, the synapse was *universalized*. Up until then no one knew whether or not the whole brain was "synaptic." Sher-rington, for one, doubted that. And most of his students and followers bypassed the question and focused exclusively on the neuromuscular junction where contiguity—and the synapse—could hardly be doubted. In 1954, however, just a few years after Jerzy Konorski (1948) and Donald Hebb (1949) published the first (and almost entirely speculative) synaptic theories of the brain, the electron physiological studies of Eduardo De Robertis, George Palade, and Sanford Palay discovered that literally all neurons are contiguous.[33] Their work made the synapse a universal fact of the nervous system and established the chemical study of the synapse—of its organization and its communication—as the science adequate to the brain.

The effect of the coming together of these three events was that out of the multitudinous past a single conceptualization of the brain emerged: the chemical, synaptic brain machine.

In the late 1950s and early 1960s neurochemistry began to emerge as *the* science of the brain, and a decade later labs all over the world were busy understanding the brain, its diseases, and its humans in synaptic and chemical terms. Brain chemicals were systematized as neurotransmitters; neurons were classified according to the neurotransmitters they produce; new tools were invented to trace chemistry-specific synaptic circuits; and soon the first links between mental and chemical processes were established, most notably with regard to memory and a relatively new disease called depression.[34]

## Memory and Synaptic Plasticity

ALREADY BY THE early 1960s research on the synaptic organization of the brain had led to a new concept of plasticity. The main author of this concept was Eric Kandel.

Beginning in the early 1960s Kandel, a young American physician and psychiatrist, was wondering whether the occasionally reported changes in the efficacy with which synapses communicate with one another could actually explain the formation of memories. Over the previous decade a combination of psycho-

logical experiments, neurosurgery, and postmortem neuropathological examinations conducted by Wilder Penfield (1891–1976), a
neurosurgeon and a student of Sherrington, and Brenda Milner, a
psychologist, were indicating that the walnut-size, horse-shaped
hippocampus is critical for memory storage. Kandel's plan was
to apply the tools of electrophysiology to single neurons of the
hippocampus and to find out if the ease with which the synapses
of these neurons trigger their impulses would change as a consequence of learning.

His animal of choice (after a brief stint with rabbits) was *Aplysia
californica*, a sea slug that was attractive because its synapses were
easily accessible and big enough so that the available electrophysi-
ological tools could be applied to them. Under the supervision
of Ladislav Tauc, working in Paris, Kandel first found that if he
ran an electrical impulse through the cell, synaptic facilitation
changed lastingly: while initially a major impulse was needed for
the cell to transmit a signal, the repeated stimulation of the cell
sensitized the cell in such a way that a minor impulse was enough
to cause transmission of a strong signal.[35]

Were such changes in facilitation the basis of memory?

Back in the United States, Kandel addressed this possibility by
studying what he called primitive forms of learning—conditioning, habituation, and sensitization—on the level of single synapses in *Aplysia*. And he found what he was looking for: what
allowed for learning, at least when it came to the gill withdrawal
reflex of the California sea slug, were changes in the efficacy with
which synapses release their action potential.[36]

The research conducted in Kandel's lab was thus largely constitutive for the emergence of something unknown prior to the late
1950s: a synaptic conception of memory.

While still working with Tauc in Paris, Kandel had begun to
refer to the variability in synaptic potentiation as synaptic plasticity.[37] By the early 1970s, especially after Tim Bliss and Terje
Lømo confirmed similar kinds of synaptic long-term potentiation

in vertebrates, synaptic plasticity emerged as one of the fastest growing areas of neuronal research.[38]

Does that mean that Kandel disproved Ramón y Cajal's observation that the adult brain is devoid of plastic changes?

Certainly not. Ramón y Cajal's concept of plasticity was an embryogenetic one: it referred to the emergence, the growth of new tissue where before there was none. Kandel's concept of plasticity, however, has been a functional one: it referred not to the growth of new cellular tissue but to the chemical processes that potentiate or depress a synaptic signal. The emergence of synaptic plasticity research since the 1960s thus hardly challenged Ramón y Cajal's observation of "essential fixity." On the contrary, the reason Kandel's concept of plasticity caused a furor was that it finally explained how a fixed and immutable brain can allow for learning: by way of synaptic plasticity.

Beginning in the 1960s one could thus claim that the brain is plastic without contradicting Ramón y Cajal's early twentieth-century observation that the adult human brain is devoid of plastic changes.

## Depression and Chemical Pathology

ALONGSIDE THE NEW concept of plasticity a new concept of pathology emerged in the work of Joseph Schildkraut (1934–2006), at the time a little-known physician at the U.S. National Institutes of Health. In 1965 Schildkraut published a review article in which he wondered whether "changes in central nervous system metabolism," that is, changes in the life-sustaining chemical reactions that happen within neurons, could be the key to understanding (at least some) cerebral diseases.[39]

The background to Schildkraut's curiosity was "pharmacological studies with . . . reserpine, amphetamine, and monoamine oxidase inhibitor antidepressants." Schildkraut's history of the past in terms of the present established a "consistent relationship between drug effects on catecholamines, especially norepinephrine, and affective or behavioral states. Those drugs which cause depletion and inactivation of norepinephrine centrally produce sedation or depression, while drugs which increase or potentiate brain norepinephrine are associated with behavioral stimulation or excitement and generally exert an antidepressant effect." Schildkraut concluded his article by establishing the "catecholamine hypothesis of affective disorders," which suggests that some if not all affective disorders "are associated with an absolute or relative deficiency of catecholamines, particularly norepinephrine, at functionally important adrenergic receptor sites in the brain," that is, at synapses.[40]

The effect of Schildkraut's review article was the emergence of a previously unknown chemical and synaptic conception of cerebral pathology. Where, up until the 1960s, cerebral diseases were almost exclusively a matter of cell death, there were now a small number of new diseases that had less to do with cell death than with neurotransmitter availability and synaptic communication: affective disorders (this included anxiety, depression, schizophrenia, and psychoses).

Yet even though the emergence of neurochemistry—of a chemical, synaptic pathology—was a significant event in the history of both neuronal research and neuronal pathology, it hardly challenged the correlation between plasticity and pathology as Ramón y Cajal (and Bichat and Bouillaud and Virchow and Remak) had brought it about. It added a new layer to and of pathology. However, Schildkraut's brain, just as Kandel's and those of their successors, was still essentially fixed. And the problem of cerebral disease was still—ultimately even for affective disorders—neurodegeneration (for it was known from postmortem pathologies that severe affective disorders result in cell death).

## (Cellular) Plasticity and Pathology (1998 to 2002)

IT WAS NOT before the year 2000, in the aftermath of the discovery of adult neurogenesis in human and nonhuman primates, that the plasticity-pathology axis that had emerged with Ramón y Cajal—via Bichat, Gall, Bouillaud, Remak, and Virchow—silently entered a period of turbulence.[41] "Silently" because initially the unexpected observation that new neurons are born in adult humans was discussed almost exclusively with respect to the one conceptual presupposition that had once made the focus on the synapse meaningful: that once maturity is reached, the central nervous system is a fixed and immutable structure, except on the level of synaptic communication. What would happen to half a century of knowledge production about synapses, synaptic communication, and synaptic networks if the brain was, after all, not fixed?[42]

However, while the often fierce discussions between those who favored a plastic conception of the brain and those who rejected it were unfolding, Fred Gage, whose lab first reported on new neurons in humans,[43] began to explore the relevance of adult neurogenesis for rethinking the diseases of the brain, specifically depression.

Over the course of the 1990s Elizabeth Gould had published a series of articles in which she showed that adult neurogenesis occurs in the hippocampus of rats, that the birth of new neurons is regulated by hormones, and that stress-induced hormones have a deleterious effect not only on older nerve cells but also on neurogenesis.[44] When, in 1998, Gage found new neurons in the hippocampi of deceased Swedish cancer patients, he wondered whether the older observation that the dentate gyrus of the hippocampus shrinks in clinically depressed patients has anything to do with Gould's observation that stress down-regulates adult neurogenesis in the dentate gyrus.

Could it be, he speculated, that the actual physiological cause of depression was not so much—as neuroscientists had believed ever since Schildkraut—an insufficient amount of neurotransmitters?

In 2000 Gage published "A Novel Theory of Depression," in which he reported—parallel to and yet independent from Jessica Mahlberg—that all of the major antidepressants his lab had studied increase the rate of new neurons in the dentate gyrus dramatically.[45] Gage used his findings to evocatively suggest that depression is not a chemical, synaptic disease—and the brain not a chemical, synaptic organ—but rather the result of a pathological down-regulation of the birth of new, yet undifferentiated, and thus still plastic neurons that would allow the organism to be adaptable to the future.

The provocation was immense. Perhaps, Gage et al. wondered, antidepressants are effective not because they augment the level of neurotransmitter and thus the potential for synaptic communication (for synaptic plasticity) but rather because they have a "neurogenic effect." Perhaps the problem of severe depression is not cell death but—cell life.

Gage's article was the first effort to attempt to think about (cerebral) pathology from the perspective of a plastic—an embryogenic—conception of the human brain.

And Prochiantz?

## Decoupling Disease from Death

A FEW WEEKS after Prochiantz and I met in his office in October 2002 to talk about his company, we met again, though this time to speak about the history of his lab. At one point, speaking about the 1980s, he told me about the experiments he had conducted "with [his] good friends Anders Bjørklund and Rusty Gage," about the survival of cultured neurons in the brains of rats.[46]

At the time all I knew about Gage was that he had discovered the birth of new neurons in the hippocampi of adult humans. When I then began to systematically read through his oeuvre and discovered the 2000 article on depression, I was struck.

Was there a link between Gage's reframing of depression and Prochiantz's company?[47]

Only many years later, after I had done fieldwork not only in Prochiantz's but also in Gage's lab (in La Jolla, California), did I learn to give a differentiated response.

Gage's provocation had been to suggest that depression is a disease of insufficient cellular plasticity, and hence of cell life. Prochiantz's provocation, however, was much more far-reaching. His ambition had been to make explicit, through his company's research, the new plasticity-pathology axis that had only been implicit in Gage's work: What if plastic (embryogenetic) processes were central not only to depression but to a vast set of cerebral diseases? If the brain were plastic—if new cells and new cellular tissue emerge throughout life—couldn't one assume that or at least explore whether the cause of cerebral diseases is not cell death but rather cell life?

Differently put, Prochiantz's ambition had been to decouple (cerebral) disease from (cell) death.

"What if cell death were normal?"

I FOLLOWED PROCHIANTZ'S short-lived company from its conception to its death. When I left Paris in summer 2003 the company had already withered. However, although Prochiantz's company never went anywhere, research on cerebral pathologies from the perspective of plasticity took off—and gave rise to a new conceptualization of disease, of what it is, of what causes it.

## (Cellular) Plasticity and (Cellular) Pathology (2002 to 2014)

IN THE DECADE after my research in Prochiantz's lab the brain became a radically different organ. At least retrospectively my departure in summer 2003 marks the end of the fixed, of the chemical, synaptic brain and the emergence of a biological, cellularly plastic organ.[48] What led to a plastic understanding of the brain, however, was only partly the rise of adult neurogenesis research, arguably the fastest growing branch of neuroscience between 2000 and 2010. At least as important, perhaps even more so, was the observation of the continuous change of cellular form in the cortex.

In 2002 the lab of Karel Svoboda showed that axons and dendrites continue to sprout and that dendritic spines get thicker and thinner or appear and disappear in the course of a single day.[49] The significance of Svoboda's work for the emergence of a plastic—and embryogenetic—conception of the adult human brain was that it decoupled the question of whether or not the adult human brain is cellularly plastic from the question of whether new neurons are born only in ancient parts of the brain such as the hippocampus or the olfactory bulb or also in distinctively human add-ons like the cortex. With Svoboda, this is to say, a more general concept of cellular (cerebral) plasticity emerged, one that could be used not only with respect to new neurons but also with respect to the continuous sprouting (and drying up) of axons and dendrites, the appearance and disappearance of spines, and the birth and death of synapses.

The perhaps most far-reaching consequence of this emergence of a more general concept of the embryogenetic plasticity of the adult brain after 2003 that I could observe to date was that what a neuron "is" gradually began to change. If from the 1950s to the 1990s neurons were chemical units—or a metabolic machine—then in the early 2000s neurons became biological, became instances of free, experience-dependent growth. And the brain became the embryogenetic organ that Prochiantz had described in the early 1990s.[50] In parallel to this embryologization of the mature central nervous system, the new conception of (cerebral) diseases that first emerged in the work of Gage and Prochiantz also got generalized.

### Disease, Differently

IT MAY SOUND strange, Gage wrote in a programmatic review article on plasticity and pathology, to study neurodegenerative diseases by focusing on embryogenetic processes. "However, the elimination of axons, dendrites and synapses is a common theme during the development of the nervous system. . . . Similarly, the . . . death of . . . cells have critical roles in brain development and maintenance in the embryonic and adult brain, and alterations in these processes are seen in neurodegenerative diseases."[51]

What Gage suggests here is that one can learn about the adult brain by studying embryogenesis precisely because (at least some) embryogenetic processes are continuing in the adult human brain. What is more, he suggests that cell death might be a normal biological event in a brain that has retained some of its embryogenetic plasticity.

Perhaps the death of cells is a regular biological event? Or, if it is not, then at least the result of a biological process gone awry? Perhaps what pathology ought to be concerned with is not cell death—the withering of neurons—but rather cell life? Living cellular processes?

Depression is today consequently far from the only disease of cell life. Over the past decade a whole series of formerly neurodegenerative diseases were reconceptualized as diseases of cell life. The perhaps most prominent of these, next to Huntington's,[52] is Alzheimer's disease. Today the problem of Alzheimer's is no longer (exclusively) that cells die. Especially in the early stages of the disease, which inflicts largely the hippocampus, they are perhaps supposed to die, as Prochiantz put it. The problem instead is that the plastic processes that keep the brain—and its human—alive, vivid, open toward the future are disrupted. As with depression, these disruptive processes are no longer conceptualized in terms of (cell) death. Rather they are now understood as living processes gone awry. The consequence is that death no longer defines the horizon of pathology. Pathology is now defined by cell life. And cell life is no limit but a vast terrain for research and therapy.[53]

"Death," as Prochiantz put it, "is a solution. Life is a problem."

# Endnotes

[1] "Couldn't it be," he summarized his work for me, "that some basic embryogenetic processes occur in the adult? For example, it could be that new nervous tissue is born: synapses, axons, dendrites, spines, even new cells. Or it could be that due to some silent embryogenesis neurons change their form. It could even be that such changes of form are ongoing, that they never cease. My experimentally grounded idea is that the continuity of embryogenetic processes in the adult render our brains plastic, and thus adaptable, throughout life."

[2] On homeotic genes, see W. J. Gehring, *Master Control Genes in Development and Evolution: The Homeobox Story*, the Terry Lectures (New Haven, CT: Yale University Press, 1998).

[3] Richard Altmann, *Über Embryonales Wachsthum* (Leipzig: private printing, 1881); Santiago Ramón y Cajal, *Histology of the Nervous System of Man and Vertebrates*, 2 vols. (Oxford: Oxford University Press, 1995).

[4] A. Jolio et al., "Antennapedia Homeobox Peptide Regulates Neural Morphogenesis," *Proceedings of the National Academy of the Sciences* 88 (1991): 1864–68.

[5] Throughout the 1990s Prochiantz had his own radio show on France Culture, on which he used to explore his ideas about brain plasticity; in addition, between 1990 and 1997, he published four popular science books that discuss the plasticity of the human brain: *Claude Bernard: La Révolution Physiologique* (Paris: PUF, 1990); *La Construction du Cerveau* (Paris: Hachette, 1993); *La Biologie dans le Boudoir* (Paris: Odile Jacob, 1995); *Les Anatomies de la Pensées: A Quoi Pense les Calamars* (Paris: Odile Jacob, 1997).

[6] Had they not been in France, where scientists are civil servants and the research budget somewhat independent from peer review, the lab would have disappeared and their research come to an end.

[7] E. Gould, P. Tanapat, and H. A. Cameron, "Adrenal Steroids Suppress Granule Cell Death in the Developing Dentate Gyrus through an NMDA Receptor-Dependent Mechanism," *Developments in Brain Research* 103 (1997): 91–93; P. S. Erikson, E. Perefilieva, T. Björk-Eriksson, A. M. Alborn, C. Nordberg, D. A. Peterson, and F. H. Gage, "Neurogenesis in the Adult Human Hippocampus," *Nature Medicine* 4, no. 11 (1998): 1313–17.

[8] Gerd Kempermann, *Adult Neurogenesis* (Oxford: Oxford University Press, 2006); Gerd Kempermann, *Adult Neurogenesis II* (Oxford: Oxford University Press, 2011). See Tobias Rees, "So Plastic a Brain: On Philosophy, Fieldwork in Philosophy, and Adult Cerebral Plasticity," *BioSocieties* 6, no. 2 (2011): 263–67.

[9] Tobias Rees, "Plastic Reason: An Anthropological Analysis of the Emergence of Adult Cerebral Plasticity in France," PhD dissertation,

University of California at Berkeley, 2006; Tobias Rees, "On Plasticity—or How the Brain Outgrew Its Histories," in *Technique, Technology, and Therapy in the Brain and Mind Sciences*, ed. Delia Gravus and Stephen Casper (New Brunswick, NJ: Rutgers University Press, forthcoming).

[10] Jean Baptiste Bouillaud, "Recherches cliniques propres à démontrer que la perte de la parole correspond à la lésion des lobules antérieurs du cerveau, et à confirmer l'opinion de M. Gall, sur le siège de l'organe du langage articulé," *Archive Général de la Médicine* 8 (1825): 25–45. As Stanley Finger points out, in the aftermath of Gall's cranioscopy there was a shift toward pathological studies of the brain. Stanley Finger, *Origins of Neuroscience* (Oxford: Oxford University Press, 1994), chapter 3.

[11] Michel Foucault, *La naissance du clinique: Une archéologie du régard médicale* (Paris: PUF, 1963).

[12] Xavier Bichat, *Traité des membranes en général et de diverses membranes en particulier* (Paris: Richard, Caille et Ravier, 1799); Xavier Bichat, *Recherches physiologiques sur la vie et la mort* (1800; Paris: Flammarion, 1994); Xavier Bichat, *Anatomie générale appliquée à la physiologie et à la médecine*, 4 vols. (Paris: Brosson, Gabon, 1801).

[13] Anatomical pathology, Bichat wrote, "consists in the examination of all the alterations our organs can undergo, at any period in which we may observe their diseases. With the exception of certain kinds of fevers and nervous affections, everything in pathology is within the province of this science." Quoted in Esmond R. Long, ed., *Selected Readings in Pathology, from Hippocrates to Virchow*, 2nd ed. (Springfield, IL: Charles C. Thomas, 1961), 88.

[14] Erwin H. Ackerknecht, *Medicine at the Paris Hospital 1794–1848* (Baltimore: Johns Hopkins University Press, 1967).

[15] To quote the perhaps most famous sentence of Bichat's *Recherches physiologiques sur la vie et la mort*: "La vie est l'ensemble des fonctions qui résiste à la mort" (57). Where death succeeded in disrupting life, or living processes, tissue degeneration—that is, disease—was the consequence.

[16] On Bouillaud, see Anne Harrington, *Medicine, Mind, and the Double Brain* (Princeton, NJ: Princeton University Press, 1989); J. D. Rolleston, "Jean Baptiste Bouillaud (1796–1881): A Pioneer in Cardiology and Neurology," *Proceedings of the Royal Society of Medicine* 24, no. 9 (1931): 1253–62; B. Stookey, "Jean Baptiste Bouillaud and Ernest Aubertin," *Journal of the American Medical Association* 184 (1963): 1024–29.

[17] Michael Hagner, *Homo Cerebralis: Der Wandel vom Seelenorgan zum Gehirn* (Berlin: Berlin Verlag, 1997).

[18] However, Bouillaud's work (he continued to study speech loss pathologically up until the 1860s) gained little attention until Paul Broca (1824–80) famously found, in the late 1850s, that several patients with language

disorders had lesions to the left frontal hemisphere. See F. Schiller, *Paul Broca: Explorer of the Brain* (Oxford: Oxford University Press, 1992).

[19] E. Clarke and L. S. Jacyna, *Nineteenth Century Origins of Neuroscientific Concepts* (Berkeley: University of California Press, 1987).

[20] Lauren Otis, *Müller's Lab* (Oxford: Oxford University Press, 2007).

[21] Norbert Kampe and Heinz Peter Schmiedebach, "Robert Remak (1815– 1865): A Case Study in Jewish Emancipation in the Mid-Nineteenth-Century German Scientific Community," *Leo Baeck Institute Yearbook* 34 (1989): 95–129; Bruno Kisch, "Forgotten Leaders in Modern Medicine: Valentin, Gruby, Remak, Auerbach," *Transactions of the American Philosophical Society* 44 (1954): 227–96.

[22] See Clarke and Jacyna, *Nineteenth Century Origins of Neuroscientific Concepts.*

[23] Theodor Meynert, *Der Bau der Großhirnrinde und seine örtliche Verschiedenheiten nebst einem pathologisch-anatomischen Korollarium* (Leipzig: Engelmann, 1868); Jean-Martin Charcot, *Leçons sur les maladies du système nerveux* (Paris: Adrien Delahaye, 1872–73).

[24] A side effect of the emergence of histology was a shift from the clinical to the cellular. Most histologists, while not uninterested in pathology, were less concerned with understanding the cellular correlates of diseases in the tradition of Bichat than with understanding the cellular composition of the brain as such.

[25] Gordon Shepherd, *Foundations of the Neuron Doctrine* (Oxford: Oxford University Press, 1991).

[26] Ramón y Cajal, *Histology of the Nervous System of Man and Vertebrates.*

[27] What led Ramón y Cajal to this argument was the observation (made previously by Richard Altmann in 1881) that neurogenesis is an event of embryogenesis exclusively. Ramón y Cajal pointed out that if new neurons do not appear after birth, then the growth of the fine structure was the only plausible way to make sense of phenomena as learning or memory.

[28] Santiago Ramón y Cajal, "La Fine Structure des Centres Nerveux," *Proceedings of the Royal Society of London,* no. 55 (1894): 444–68.

[29] Santiago Ramón y Cajal, *Regeneration and Degeneration of the Nervous System,* ed. and trans. Raoul M. May (Oxford: Oxford University Press, 1928), 509, 750.

[30] What is remarkable is that almost all of the different brains that flourished all over Europe presupposed that humans are born with a definite number of neurons and that once adulthood is reached, the brain is a somewhat immutable cellular tissue. The work of Sherrington and Adrian—their effort to map how electric information travels—made sense only if the actual neuronal wires were stable; the work of the cytoarchi-

tects made sense only if the neuronal patterns were unchanging. Hardly any one of these brains thus challenged Ramón y Cajal's idea of "essential fixity."

[31] The one local tradition for which the synapse—or at least nerve impulse transmission—was the central concept for understanding the brain thus began to spread its web across the Western world, just at the time when the two other major lines of research ceased (Germany and Spain).

[32] Here is a brief description of the technical experiments that established the chemical quality of the nerve cell. In 1939, independently of the older pharmacological studies of nervous tissue, Alan Hodgkin and Andrew Huxley, students of Edgar Adrian at Cambridge who were experimenting on the giant axon of the Atlantic squid, succeeded in recording how the release of an electronic impulse changed the electrical charge of the "conducting fluid" outside of the axon. When, toward the end of World War II, they returned to the bench and continued their work, they could document that the ionic current in a nerve cell during impulse depended on two phases of cell membrane permeability: rise of impulse needs sodium permeability and fall of impulse needs potassium permeability. These insights led them to formulate the ionic theory of transmission, proposing that transmission of impulses is a process that involves receptors controlled by ions. See A. L. Hodgkin and A. F. Huxley, "Resting and Action Potentials in Single Neurons," *Journal of Physiology* 104 (1945): 176–95; A. L. Hodgkin and A. F. Huxley, "Potassium Leakage from an Active Nerve Fibre," *Journal of Physiology* 106 (1947): 341–67. Strikingly for students of Lucas and Adrian, they did not use the term *synapse* in their work. Amid the war, however, Wilhelm Feldberg, a German refugee who had worked with Henri Dale on acetylcholine, and Alfred Fessard, who had worked with Adrian in Cambridge, documented that the peripheral nerves of the electric ray (Torpedo) released a chemical called eserine. See Wilhelm Feldberg and Alfred Fessard, "The Cholinergic Nature of the Nerves of the Electric Organ of the Torpedo (Torpedo Marmorata)," *Journal of Physiology* 101 (1942): 200–215.

Was transmission a chemical event? John Eccles, Sherrington's last student, disagreed. In part building on the work of Hodgkin and Huxley, Eccles provided evidence—his focus was on the neuromuscular junction in frogs—for how synapses integrate and coordinate nervous impulses. Analyzing intercellular recordings with mini-electrodes, he could show that if the arriving impulse is connected to what he called "excitatory synapses of the cell," the excitability of the cell increases, and when the "inhibitory synapses" make the cell respond, the result is a diminution of excitability. See John Eccles, "Conduct and Synaptic Transmission in the Nervous System," *Annual Reviews in Physiology* 10 (1948): 93–116.

At this point Bernhard Katz (like Feldberg, he had fled Nazi Germany) disagreed with Eccles. In 1948 Katz, who had been working under Sherrington, had collaborated closely with Eccles, and was also focusing on the neuromuscular junction, established that nervous communication is not an electrical process. (His focus too was on frogs and also on crustacea.) In a critique of Hodgkin and Huxley he showed that nervous communication is a chemical process, that it occurs by way of a release of chemicals present in a cell, and that the synapses is in fact a chemical entity. To be more precise, he showed how acetylcholine causes a large and very brief increase in the ionic permeability of what he called the "synaptic membrane." The consequent ionic influx across the membrane causes the actual electrical potential transmitted "between synapses." Bernhard Katz, "The Electric Properties of the Muscle Fibre Membrane," *Proceedings of the Royal Society, B* 135 (1948): 506–34. This work was refined by Paul Fatt and Bernhard Katz, "An Analysis of the End-Plate Potential Recorded with an Intra-cellular Electrode," *Journal of Physiology* 115 (1952): 320–70. In 1952, then, Eccles repeated the work of Fatt and Katz and applied it to motor neurons—and was convinced that all transmission is chemical.

[33]  Eduardo De Robertis and H. Bennet, "Submicroscopic Vesicular Component in the Synapse," *Federation Proceedings* 13 (1954): 35; Eduardo De Robertis and H. Bennet, "A Submicroscopic Vesicular Component of Schwann Cells and Nerve Satellite Cells," *Experimental Cell Research* 6 (1954): 543–45; George Palade, "Electron Microscope Observations of Interneuronal and Neuromuscular Synapses," *Anatomical Record* 118 (1954): 335–36; Sanford Palay, "Synapses in the Central Nervous System," *Journal of Biophysics, Biochemistry, and Cytology* 2, supplement (1956): 193–202. Toward the end of the 1940s Jerzy Konorski, *Conditioned Reflexes and Neuron Organization* (Cambridge: Cambridge University Press, 1948) and Donald Hebb, *The Organization of Behavior* (New York: Wiley, 1949) published the first theories of the brain as a synaptic organ. They suggested—independently from one another—that the entire nervous system is organized in synaptic, function-specific circuits, which are governed by synaptic communication. What led them to this suggestion was the work that had given rise to Ramón y Cajal from the perspective of then cutting-edge synaptology. If neurogenesis is an event of embryogenesis exclusively, they argued, and if the adult brain is an immutable structure, then synaptic communication was the only dynamic element of the nervous system—and hence the only plausible candidate for understanding the brain.

[34]  For a review of these processes, see Gordon Shepherd, *Creating Modern Neuroscience: The Revolutionary 1950s* (Oxford: Oxford University Press, 2010).

[35]  L. Tauc and E. R. Kandel, "An Anomalous Form of Rectification in a Molluscan Central Neurone," *Nature* 202 (1964): 145–47; E. R. Kandel and L. Tauc, "Heterosynaptic Facilitation in Neurones of the Abdominal Ganglion of Aplysia depilans," *Journal of Physiology* 181 (1965): 1–27; E. R. Kandel and L. Tauc, "Mechanism of Heterosynaptic Facilitation in

the Giant Cell of the Abdominal Ganglion of Aplysia depilans," *Journal of Physiology* 181 (1965): 28–47.

36  V. Castelucci et al., "Neuronal Mechanisms of Habituation and Dishabituation of the Gill-Withdrawal Reflex in Aplysia," *Science* 167 (1970): 1745–48.

37  See Kandel and Tauc, "An Anomalous Form of Rectification in a Molluscan Central Neurone." For a prehistory of this concept of plasticity with respect to potentiation, see Rees, "Plastic Reason."

38  T. V. P. Bliss and T. Lømo, "Long-lasting Potentiation of Synaptic Transmission in the Dentate Area of the Anaesthetized Rabbit Following Stimulation of the Perforant Path," *Journal of Physiology* 223 (1973): 331–56. See as well Gordon Shepherd, *The Synaptic Organization of the Brain* (Oxford: Oxford University Press, 1979); Shepherd, *Creating Modern Neuroscience*.

39  On the rise of the "the neuron as a metabolic unit," that is, of the chemical comprehension of neurons, since the mid-1950s, see Shepherd, *Foundations of the Neuron Doctrine*, 286–89.

40  Joseph Schildkraut, "The Catecholamine Hypothesis of Affective Disorders," *American Journal of Psychiatry* 122 (1965): 509–22.

41  Erikson et al., "Neurogenesis in the Adult Human Hippocampus"; E. Gould et al., "Neurogenesis in the Dentate Gyrus of the Adult Tree Shrew Is Regulated by Psychosocial Stress and NMDA Receptor Activation," *Journal of Neuroscience* 17 (1997): 2492–98. Adult neurogenesis research had been long in the making. First reports date back to the early 1960s (Altman) and 1970s (Kaplan). However, it was only with the observation of the birth of new neurons in humans and nonhuman primates in 1998 that adult neurogenesis began to seriously challenge the two presuppositions on which twentieth-century neuronal research had been built: that humans are born with a definite number of neurons and that once adulthood is reached the spectacular growth of the fine structure comes to a definite end and the brain becomes a fixed and immutable cellular tissue. For a history of adult neurogenesis research, see especially Kempermann, *Adult Neurogenesis* and *Adult Neurogenesis II*.

42  For a discussion of the history of adult neurogenesis research see Kempermann, *Adult Neurogenesis* and *Adult Neurogenesis II*.

43  Erikson et al., "Neurogenesis in the Adult Human Hippocampus."

44  H. A. Cameron et al., "Differentiation of Newly Born Neurons and Glia in the Dentate Gyrus of the Adult Rat," *Neuroscience* 56 (1993): 337–44; E. Gould et al., "Blockade of NMDA Receptors Increases Cell Death and Birth in the Developing Rat Dentate Gyrus," *Journal of Comparative Neurology* 340 (1994): 551–65; E. Gould et al., "Adrenal Steroids Suppress Granule Cell Death in the Developing Dentate Gyrus through an NMDA Receptor-Dependent Mechanism."; E. Gould et al., "Neurogenesis in the

Dentate Gyrus of the Adult Tree Shrew Is Regulated by Psychosocial Stress and NMDA Receptor Activation."

45  J. E. Mahlberg, "Chronic Antidepressant Treatment Increases Neurogenesis in Adult Rat Hippocampus," *Journal of Neuroscience* 20, no. 24 (2000): 9104–10; B. L. Jacobs, H. van Praag, and F. Gage, "Adult Brain Neurogenesis and Psychiatry: A Novel Theory of Depression," *Molecular Psychiatry* 5 (2000): 262–69.

46  P. Brindon et al., "Survival of Intracerebrally Grafted Rat Dopamine Neurons Previously Cultured in Vitro," *Neuroscience Letters* 61 (1985): 79–84.

47  "Look," Prochiantz replied when I returned with my discovery, "Rusty and I have known each other for almost twenty years. We are in a steady, ongoing conversation about the brain, about plasticity, about the idea of thinking about the adult brain in terms of an ongoing embryogenesis. We are even about to edit a book together about neural stem cells."

48  I write these lines retrospectively. When I did fieldwork in 2002–3 no one could know yet whether the new ways of conceptualizing the brain were tentative, short-lived explorations or would grow into a somewhat stable way of understanding the brain.

49  J. T. Trachtenberg et al., "Long-term *in vivo* Imaging of Experience-Dependent Synaptic Plasticity in Adult Cortex," *Nature* 420 (2002): 788–94; Karen Zito and Karel Svoboda, "Activity-Dependent Synaptogenesis in the Adult Mammalian Cortex," *Neuron* 35 (2002): 1015–17. For a review, see A. Holtmaat and Karel Svoboda, "Experience-Dependent Structural Synaptic Plasticity in the Mammalian Brain," *Nature Reviews Neuroscience* 10 (2009): 647–58.

50  Tobias Rees, "Being Neurologically Human Today: Life, Science, and Adult Cerebral Plasticity (An Ethical Analysis)," *American Ethnologist* 37, no. 1 (2010): 150–66; Rees, "On Plasticity." For a review, see Douglas Field, *Beyond the Synapse: Cell-Cell Signaling in Synaptic Plasticity* (Cambridge: Cambridge University Press, 2012).

51  B. Winner, Z. Kohl, and F. H. Gage, "Neurodegenerative Disease and Adult Neurogenesis," *European Journal of Neuroscience* 33, no. 6 (2011): 1139–51.

52  A. Ernst et al., "Neurogenesis in the Striatum of the Adult Human Brain," *Cell* 156 (2014): 1072–83.

53  For a comprehensive review, see Y. Mu and F. H. Gage, "Adult Hippocampal Neurogenesis and Its Role in Alzheimer's Disease," *Molecular Neurodegeneration* 85 (2011): 1–9. For more recent overviews, see C. P. Fitzsimons et al., "Epigenetic Regulation of Adult Neural Stem Cells: Implications for Alzheimer's Disease," *Molecular Neurodegeneration* 9 (2014): 1–29; N. C. Inestrosa and L. Varella-Nallar, "Wnt Signaling in the Nervous System and in Alzheimer's Disease," *Journal of Molecular Cell*

*Biology* 6 (2014): 64–74. I would like to specify: the emergence of a plastic brain and a concern with plastic, with living cellular processes is qualitatively vastly different from older experiments in which cells were taken from a donor, cultured in a Petri dish, and then added to the diseased brain region of a patient. It is qualitatively different because these older efforts assumed the brain to be fixed and immutable.

# Index

Abelson, Robert, 251, 253
Adorno, Theodor W., 39–40, 45, 188
agency, 61, 178, 222, 285
Alzheimer's disease, 333
antidepressants, 327–28, 329–30
aphasia, 112–51; Bouillaud on, 316;
children and, 207; Goldstein
on, 117–21, 128, 132–33, 150,
167–68; Head and Goldstein on
previous aphasiology, 113, 121,
123; Head on, 117, 118, 121–32;
Jackson on, 115, 120, 121, 128;
localization theories in study
of, 175; Luria's work on, 172; of
Zasetsky, 160, 164, 166, 183, 185,
187
*Aphasia and Kindred Disorders of
Speech* (Head), 114, 117, 118,
121–23
artificial intelligence (AI): goals of
early research, 196–97; legacy
in neuroscience and cognitive
science, 194; lineage of, 198–200;
neural plasticity's importance for,
197–98; neuroses simulated by,
249–56; Turing and, 209, 211–12
Ashby, W. Ross, 208–9, 210
Atkinson, Richard C., 254, 255
*Authoritarian Personality, The* (Adorno
et al.), 39–40, 45, 53
authority, obedience to, 45–48, 53
autism, 268, 270, 294–97

auto-affection, 26–27, 31
automaticity. *See* automatism
automatism: Descartes's automaton,
198, 199, 214; and doubling of
consciousness, 90–91, 93–94, 95,
97; forensic anxieties of mental,
82–90; of human thinking, 197,
214; Jackson on brain function,
200; and personhood, 88, 101–2;
rigidity attributed to, 206–7;
Turing on humans and automata,
211; von Neumann on, 212–13
autonomy: automaticity and, 197;
cell, 311; cognitivism on, 36;
computational rationality and,
61; Goldstein on, 148; Metzinger
denies, 22; Skinner and, 52, 59

behaviorism: authoritarianism
attributed to, 40–41, 48;
cognitive scientists repudiate, 60;
introspective methods rejected
by, 14; on meaning, 41–42; and
Milgram's obedience to authority
experiment, 45; Miller, Galanter,
and Pribram on, 241–42, 248,
256; Orne's criticism of, 15–16;
relinquishes its larger social
ambitions, 59; of Skinner, 37–39,
50–52
behavior modification, 48, 58
Benedikt, Moritz, 78

Benschop, Ruth, 9–11
Bertalanffy, Ludwig von, 207, 213
Bichat, Xavier, 315–16, 318, 322,
   328, 329
Binet, Alfred, 89, 93–94
Bouillaud, Jean Baptiste, 315–17,
   323, 328, 329
brain, the: automaticity of, 197;
   biological, cellularly plastic, 331–
   32; Broca's area, 287; chemical,
   synaptic, 323–25, 331; circuits
   attributed to, 220–22; computers
   as metaphor for, 219–20, 222,
   227, 229, 231, 235, 248, 254, 258;
   computers model animal brains
   with synaptic architectures,
   195–96; cybernetics and brain
   science, 209; Descartes on, 199;
   diagramming circuits of, 219–33;
   as digital machine, 194–98,
   214; doesn't have a brain, 247;
   embryogenetic plasticity of
   adult, 312–13, 330, 332, 333;
   emergence of cerebral pathology,
   315–18, 329; fixity attributed to,
   312, 313, 319, 322, 327; Heath's
   brain stimulation experiments,
   35–36, 43–45, 54–56, 58;
   hippocampus, 326, 329, 330,
   332, 333; histological research,
   1897–1914, 318–22; how adult
   cerebral plasticity research has
   decoupled pathology from death,
   309–33; intelligence and size of,
   166; Jackson on, 67, 68, 95, 200;
   limits between consciousness
   and, 32–33; Luria's dialectical
   understanding of, 165, 176; the
   normal versus the pathological,
   70–72; pleasure centers, 35,
   39, 45, 48, 50; processes in
   construction of consciousness,
   23–24; sensory-motor cortex,
   271–75; Sherrington on nervous
   system integration, 114–16;
   Sherrington's cat decortication

experiments, 130; as transparent
   from standpoint of subject,
   25–26; Turing on computers
   modeling, 210; von Neumann
   on, 212. *See also* brain injuries;
   localization theories; neurons;
   synapses
brain injuries: Goldstein on, 117,
   118, 121, 134, 136, 137, 144, 147,
   150, 167–68; Head on, 117, 121,
   123, 125, 126, 147; James on,
   201; Luria on, 172–75; as method
   for exploring brain, 26; Peirce on,
   201–2; reorganization in response
   to, 202–4; Schneider case, 117,
   132–33; somatic integration in
   response to, 113; of Zasetsky,
   159–60, 182
Broca, Paul, 127, 166, 317, 323
Broca's area, 287
Burks, Arthur W., 237
Butler, Judith, 30

Cambridge Anthropological
   Expedition to Torres Straits
   Islands, 11–12, 13, 14, 15
Cambridge Laboratory of
   Experimental Psychology, 12–13
Canguilhem, Georges, 72, 207, 208
Carpenter, William, 84
cells: autonomy of, 311; death of,
   313–14, 317, 322, 328, 330,
   333; in emergence of cerebral
   pathology, 315–18; plasticity of,
   312. *See also* neurons
central nervous system: fixity
   attributed to, 329; Sherrington
   on nervous system integration,
   114–16. *See also* brain, the
Charcot, Jean-Martin, 318, 323
Chomsky, Noam, 41–42, 52–53, 60,
   61
Claparède, Édouard, 205
cognitivism: of Chomsky, 53; versus
   continental philosophy, 21–22,
   28; free will and, 25; Harvard

Center for Cognitive Studies, 42; Miller, Galanter, and Pribram on, 241–42; moral freedom identified with, 59

Colby, Kenneth Mark, 250–51, 252

computational rationality, 60–61

computing: conceptual connections between computers and neurophysiology, 194; evolution of, 205–13; modeling animal brains with synaptic architectures, 195–96. *See also* digital computers

consciousness: as already ongoing, 247; brain processes in construction of, 23–24; doubling of, 90–94, 95–100; epilepsy and total loss of, 88–90; Head on, 120, 131, 146; Jackson on object and subject, 100–101; limits between the brain and, 32–33; Luria on, 179; personal identity grounded in, 66; subjectivity coincides with, 22

continental philosophy, 21–22, 24, 31–32, 33

creativity, 31, 198, 204, 211

Crichton-Browne, James, 69, 91–92

cybernetics, 197, 207–9, 224, 231, 240, 322

Damasio, Antonio, 26

Daston, Lorraine, 1–2, 5, 9, 15

death, 309–33; cell, 313–14, 317, 322, 328, 330, 333

depression, 325, 327–28, 329–30, 331, 333

Descartes, René, 70, 198–99, 214

digital computers: as awaiting their determination, 213–14; imitating any other discrete-state machine, 196–97, 206, 209–10; as metaphor for the brain, 219–20, 222, 227, 229, 231, 235, 248, 254, 258; neural plasticity concept as influence on, 198; simulating

human intelligence, 210; Turing on universal, 197, 206; von Neumann on, 212–13

Echeverria, Manuel, 82, 87, 88

embryogenesis, 310, 312–13, 319, 327, 330, 332–33

emotions, 269, 289–91, 292–93, 297–99

empathy, 269, 270, 288–91, 293–94, 298

Enlightenment (*Aufklärung*), 21, 24, 29, 31, 33, 51, 61

epilepsy: *grand mal* and *petit mal* episodes, 79–80; Jackson on, 72, 73–76, 77, 79, 82, 83–85, 86, 87, 91; masked (larvated), 80–82, 83, 85, 87; Maudsley on, 76, 78, 79, 81–82, 87, 89; and mental automatism, 82–90; as normalized abnormalities, 72–77; and total loss of consciousness, 88–90; violence associated with, 77–79, 83

equipotentiality, 203, 213

Esquirol, J. E. D., 77, 79–80, 82, 89

existential devices, 224, 249, 251, 254, 257, 258

experience: border of the transcendental and, 32; in computational developmental neuroscience, 195; of immediacy of consciousness, 23; of myself, 24; sensory, 13; subject's eliminated from psychology, 5, 6, 9

experimental method: subject stability as goal of, 2, 4, 6–9, 15–16; subject training as routine part of, 5

experimental psychology, ethnology of, 1–16

Fairet, Jules, 78–79, 86, 87

Ferrier, David, 85

flowcharts: in cognitive psychology,

253–54; Goldstine and von
Neumann flow diagrams,
233–40, 242, 246, 257; in Miller,
Galanter, and Pribram's *Plans
and the Structure of Behavior,*
240–49; origin of, 222–23; for
representing brain circuits,
220–22; variety of meanings of,
256–58
Foucault, Michel, 20, 21, 22, 24–25,
28–30, 32, 81
free will, 25, 33, 36, 51, 59, 60, 61

Gage, Fred, 329–31, 332
Galanter, Eugene, 240–49, 254, 256,
257
Galison, Peter, 1–2, 5, 9, 15
Gall, Franz Joseph, 68, 71–72, 120,
316, 329
Gallese, Vittorio, 269, 289–91, 298
Gaub, Jerome, 71
Gelb, Adhémar, 117, 118, 122,
132–33, 138
Gestalt psychology, 115, 118, 203,
204, 205
Goldstein, Kurt, 132–45; on
abstract attitude, 144, 147, 149,
150; on aphasia, 117–21, 128,
132–33, 167–68; on atomistic
symptomatology, 133–36; on
catastrophe reaction, 146, 147,
148, 204, 205; on categorial
behavior, 136–37, 146–47; ear
puff experiments, 143–44, *144*;
grasping and pointing tests of,
142–43, *143*; Head contrasted
with, 138, 145–51; Luria
influenced by, 169, 170, 172;
on neural plasticity, 203–4; on
normality and pathology, 72;
on norms, 121, 133, 147–48;
*The Organism,* 114, 117, 118,
133–35, 138, 139–40, 146, 148;
on preferred behavior, 139–42;
on previous aphasiologists, 113,
167; on reflex theory, 135–36;

relationship with Head, 117–18,
122; Schneider case, 117, 132–33;
theory of wholeness of, 114, 133,
135, 137–39, 145, 148, 150, 168,
203; on tonus, 133, 138–44, 149
Goldstine, Herman Heine, 233–40,
242, 246, 257
Gould, Elizabeth, 329
Gowers, William, 80, 81, 82, 83,
85–86, 89

Head, Henry: on aphasia, 117, 118,
119–32, 150; *Aphasia and Kindred
Disorders of Speech,* 114, 117,
118, 121–23; on brain injuries,
117, 121; collaboration with
Rivers, 118–19, *119,* 130; on
consciousness, 120, 131, 146; on
disorders of symbolic thought
and expression, 123, 130; on
freedom, 123, 148; Goldstein
contrasted with, 138, 145–51;
on language, 121, 126–28,
147, 149–50; as man of letters,
116–17; monism of, 150–51; on
organismic integration, 114,
123–26, 129, 130–32, 149; on
previous aphasiologists, 113, 121,
123; relationship with Goldstein,
117–18, 122; on vigilance,
128–31, 139, 145–46, 147, 149–50
Heath, Robert Galbraith: attempted
control of operant behavior with
intracranial self-stimulation,
48–50; and Kennedy hearings
on neuroscience, 58, 59–60;
reparative therapy experiment
of, 35–36, 54–56; schizophrenia
experiment of, 42–45
Hebb, Donald, 27, 160, 195, 325
hedonist psychology, 35–38, 42
heterosexuality, 35–36, 54–56
hippocampus, 326, 329, 330, 332,
333
histology, 318–22
homeoproteins, 310–12

homeostasis, 209, 243
homosexuality, 35–36, 54–56, 57
human experimentation, 57–58

incongruence: autism and, 294, 296; as cognitive, 298; empathy and, 288–91; language and, 287–88; in mirror neurons, 270, 271, 278–81, 282; psychopathy and, 297
inner-ear lesions, 143–44
insanity: degrees of, 98–99; epileptic, 78–79, 82, 83, 85, 87, 88, 89, 94–95, 97, 98; Jackson on, 67–70, 74–76, 97–99; personhood and, 65–70
insight, 201, 203, 204–5
intelligence: intelligent automata, 214; intelligent machines, 209–11; Köhler on, 204–5. *See also* artificial intelligence (AI)
introspection: elimination from experimental setting, 4–5, 14, 15; Titchener's use of, 11, 13–14; in Wundt's psychology laboratory, 9, 11, 13, 14
intuition, 201, 203

Jackson, John Hughlings: on aphasia, 115, 120, 121, 128; on automatism, 83–84, 91; on the brain, 67, 68, 95, 200; on epilepsy, 72, 73–76, 77, 79, 82–87, 91; "The Factors of Insanities," 67, 98–100; general theory of neurological disorders, 74–76; on insanity, 67–70, 97–99; on intuition, 201; on mental diplopia, 95–100; Mitchell's *Lectures on Diseases of the Nervous System* dedicated to, 92; on nervous system, 67, 73–74, 86, 95, 115; on new person in pathological states, 99–102; on personal identity and insanity, 67–70, 94–103; on play of the mind, 96

James, William, 75, 84, 200–201, 208, 242, 247

Kandel, Eric, 61, 161, 325–27, 328
Kant, Immanuel, 21, 22, 24, 26–27, 29, 32
Koffka, Kurt, 139, 205
Köhler, Wolfgang, 204–5, 212

Lacan, Jacques, 24, 231–33, 271
language: Chomsky on, 41–42, 60; Head on, 121, 126–28, 147, 149–50; mirror neurons and, 268, 269, 270, 287–88. *See also* aphasia
Lashley, Karl, 41, 60, 113, 175, 202–3, 212, 213
Latour, Bruno, 8
Laycock, Thomas, 80, 84
LeDoux, Joseph, 21, 22, 28
Leys, Ruth, 269, 291
localization theories: Broca on, 166; Gall on, 316–17; Goldstein on, 118, 120, 132, 133, 134; Head on, 118, 120, 127, *128, 129*; Lashley on, 202; Luria on, 175–76; Monakow on, 202; von Neumann on, 212
Locke, John, 65–66, 91
logic machines, 199–200
Lombroso, Cesare, 78
Luria, Alexander, 159–89; on the brain, 165, 176; on brain injuries, 172–75; in child psychology, 179–80; denunciations of work of, 178–79; Goldstein as influence on, 169, 170, 172; *The Man with a Shattered World*, 160, 162–63, 170, 180–89; on neural plasticity, 164, 170, 174, 175, 177–79, 188–89; Russian Revolutionary context of, 176–77; Zasetsky case, 160, 164, 180–89

Malabou, Catherine: on the aphasic novel, 183; on clinicians shaping

the paradigm, 169; on destructive
plasticity, 162, 163–64, 187; on
Luria's account of Zasetsky, 160,
186; on machinic understandings
of brain, 195; on neuroplasticity
as explosiveness, 61; on plasticity
and capitalism, 160, 162; on
plasticity and individual identity,
161
maladaptive sexuality theory, 54
Maudsley, Henry: on automatism,
86, 89; on childhood insanity,
85; on epilepsy, 76, 78, 79, 81–82,
87, 89
McCulloch, Warren, 224–33, 251,
257
McCulloch-Pitts logical neuron
diagrams, 224–33
meaning, 41–42, 60
memory, 228, 233–34, 247, 254,
255, 325–27
Metzinger, Thomas, 22–23, 25, 30,
31, 32
Meynart, Theodor, 318, 323
Milgram, Stanley, 45–48, 53
Miller, George, 240–49, 254, 256,
257
mind reading, 298
mirror neurons, 268–99; autism
and, 268, 270, 294–97;
emergence as research program,
271–75; empathy and, 269, 270,
288–91; high order mirroring
and nonsimultaneity, 282–86;
incongruence and, 270, 271,
278–81, 282; language and,
268, 269, 270, 287–88;
nonsimultaneity and, 270, 271,
275–78, 281, 282; psychopathy
and, 270, 291–94
Mitchell, Silas Weir, 92–93
Monakow, Constantin von, 122,
202, 203
Myers, Charles S., 12–13, 118

nervous system: chips that

mimic neural systems, 196;
Descartes on, 198–99; digital
representations of, 195; Head's
general theory of nervous
functioning, 117; Jackson on,
67, 73–76, 86, 95, 115; neurons
that fire together wire together,
33, 195; seen as fixed, 311;
Sherrington on integration of,
114–16, 145, 147, 208; Turing on
computers modeling, 210; von
Neumann on, 212, 213. *See also*
brain, the; neurons
neural nets, 224–33
neural (synaptic) plasticity, 28;
automaticity of thinking based on,
214; Canguilhem on, 207; cellular,
328–32; cold war politics and
limitation of, 60; contemporary
notions of, 166; with coterminous
stimuli, 33; Descartes on, 199;
destructive, 160, 162, 163–64,
187, 188; embryogenetic versus
functional concepts of, 327;
Goldstein on, 203–4; Heath's
commitment to, 35; historical
approach to, 165; importance for
artificial intelligence, 197–98;
James on, 200–201, 208; Lashley
on, 202–3; in link between
neuronal matter and psychological
function, 170–71; Luria on,
164, 170, 174, 175, 177, 178–79,
188–89; memory and, 325–27;
new concepts of 1960s, 322–23;
plasticity-pathology axis, 322, 329,
330; political aspect attributed to,
160–64; precludes easy reduction
of cognition to automatic
machinery, 205–6; Ramón y Cajal
on, 320, 321, 322; reparative,
166–67, 188; in self-fashioning,
30–31; submolecular mechanisms
of, 61; and technological
conceptualizations of brain, 195
neural singularity, 31

101–2; and pathology, 65–70, 90, 94–103; preconditions of, 100–101. *See also* personal identity

Pitts, Walter, 224–25, 228, 233

plasticity: of adult brain, 312–13; behavioral, 36, 60; in computing technologies, 197; cybernetic, 209; how adult cerebral plasticity research has decoupled pathology from death, 309–33; of human beings, 2, 177–78, 179–80; of machines, 208–9; narrated, 188; of reason, 33; von Neumann on error and, 213. *See also* neural (synaptic) plasticity

pleasure: in Heath's reorientation experiment, 35–36, 54–56, 58; in operant conditioning, 36–39; pleasure centers of the brain, 35, 39, 45, 48, 50; role in behavior, 48–50

Polanyi, Michael, 207, 208

practice effect, 3–4, 14

preference, circular, 228–29

Pribram, Karl, 240–49, 254, 256, 257

Prochiantz, Alain, 309–14, 330–31, 332, 333

psychology: elevating to status of hard science, 39; Gestalt, 115, 118, 203, 204, 205; hedonist, 35–38, 42; synthesis of neuroscience and, 200. *See also* behaviorism; cognitivism; experimental psychology, ethnology of

psychopathy, 270, 291–94

Rado, Sandor, 42, 43, 54, 56

Ramachandran, V. S., 269, 282–83, 295, 296, 298

Ramón y Cajal, Santiago, 27, 311, 319–22, 323, 327, 328, 329

reason, 29–30, 33, 61, 199

reflex arcs, 37, 115, 116, 136, 243

reflexes, 115, 134, 135–36, 139, 245

reinforcement, 37–39, 51–52

Remak, Robert, 317, 318, 322, 328, 329

reparative theory, 35–36, 54–56

responsibility, 51, 57, 77

reward and punishment: in Heath's brain stimulation experiments, 35–36, 45, 49–50; in Milgram's obedience to authority experiment, 46–48; in operant conditioning, 36–39; Skinner on, 59

Rivers, W. H. R., 11–12, 118–19, *119,* 130

Rizzolatti, Giacomo, 268, 269, 272–80, 287–91, 298

Sacks, Oliver, 162–63, 169–70, 183

Schildkraut, Joseph, 327–28, 329

schizophrenia, 42–45, 48–50, 251

Schneider, Johann, 117, 132–33

Schwann, Theodor, 317, 318

self, the: denial of, 22–24, 31; doubling of, 90–94; Foucault's concepts of technologies of, 30; fragmentary, 271; Head and Goldstein on, 119–21; pathological dissolution of, 70; romanticized concepts of, 197; self-transformability, 32. *See also* personal identity

self-actualization, 121, 145, 148–49, 150

self-fashioning, 30–31

sexual orientation disturbance (SOD), 56

Sherrington, Charles: on brain as electrical organ, 323, 324; Goldstein's critique of, 135–36; and Head's concept of vigilance, 129–30; Jackson influenced by, 120; on nervous system integration, 114–16, 145, 147, 208; on synapses, 27, 323, 324–25

Shiffrin, Richard M., 254, 255

Simondson, Gilbert, 213, 214
Skinner, Burrhus Frederic, 37–39;
  *Beyond Freedom and Dignity,*
  51–53; Chomsky's criticisms of,
  41–42, 52–53, 60; and Kennedy
  hearings on neuroscience, 57,
  58–59; on mind control, 48; *On
  Behaviorism,* 54; public hostility
  toward, 53–54; and split in
  Harvard psychology department,
  40–41; *Verbal Behavior,* 41, 52,
  60; *Walden Two,* 38, 50
socialist realism, 181, 182, 188
subconscious, the, 93, 94
subject (experimental): experience
  eliminated from psychology, 5,
  6, 9; in Heath's brain stimulation
  experiments, 50; stability of, 2,
  4, 6–9, 15–16; standardization
  of, 10; Titchener on, 13–14;
  training, 1–2, 5, 9, 15; in Wundt's
  laboratory, 9–11, 15
subjectivity: artificial, 31; brain as
  transparent from standpoint
  of subject, 25–26; contention
  regarding, 5, 14; denial of, 22–24,
  30; in experimental psychology,
  1–2; as its own formation and
  transformation, 28–29; models
  that go no further than central
  nervous system, 164; subjective
  site of, 26–27; of Zasetsky, 160
Svoboda, Karel, 332
synapses, 27–32; in affective
  disorders, 328; chemical nature
  of communication of, 322,
  324; chemical, synaptic brain
  emerges, 323–25; neurosynaptic
  chips, 196; as strengthened
  with use, 195. *See also* neural
  (synaptic) plasticity
synaptic delay, 225

Tauc, Ladislav, 326

test-enhanced learning, 3
Test-Operate-Test-Exit (TOTE) unit,
  243–46, *244*
Thorndike, Edward, 36–37
time, 229, 232, 235, 239
Titchener, Edward, 11, 13–14
tonus, 133, 138–44, 149
*Tonus* (film), 141–43, *142, 144*
Torres Straits expedition of 1898,
  11–12, 13, 14, 15, 118
trained judgment, 1–2, 5, 9, 15
Trousseau, Armand, 86–87
Tulane electrical brain stimulation
  program, 42, 43–45
Turing, Alan: artificial intelligence
  and, 209, 211–12; on computers
  imitating any other discrete-
  state machine, 197, 206, 209–10;
  flow diagrams of, 233, 239,
  *240*; on machines competing
  with humans, 210–11; Turing
  machine, 206, 225, 234

vigilance, 128–31, 139, 145–46, 147,
  149–50
Virchow, Rudolf, 317–18, 322, 328,
  329
von Neumann, John, 212–13,
  233–40, 242, 246, 257

*Walden Two* (Skinner), 38, 50
Watson, John, 37, 38
"What Is Enlightenment?"
  (Foucault), 21, 28
Whytt, Robert, 71
Wicker, Bruno, 269, 289–91, 297–98
Wundt, Wilhelm, 9–11, 12, 13, 14

Zasetsky, L., 180–89; aphasia
  of, 160, 164, 166, 183, 185,
  187; brain injury of, 159–60,
  182; faces the future, 186–88;
  surrealist images of, 183–84
Žižek, Slavoj, 23, 24, 31

Berkeley Forum in the Humanities

Teresa Stojkov (ed.), *Critical Views: Essays on the Humanities and the Arts*

J. M. Bernstein, Claudia Brodsky, Anthony J. Cascardi, Thierry de Duve, Ales Erjavec, Robert Kaufman, and Fred Rush, *Art and Aesthetics after Adorno*

Malcolm Bull, Anthony J. Cascardi, and T. J. Clark, *Nietzsche's Negative Ecologies*

David Bates and Nima Bassiri (eds.), *Plasticity and Pathology: On the Formation of the Neural Subject*